Lecture Notes in Computer Science   13361

More information about this series at https://link.springer.com/bookseries/558

Laura Kovács · Karl Meinke (Eds.)

Tests and Proofs

16th International Conference, TAP 2022
Held as Part of STAF 2022
Nantes, France, July 5, 2022
Proceedings

 Springer

Editors
Laura Kovács ⓘ
Faculty of Informatics
TU Wien
Vienna, Austria

Karl Meinke ⓘ
Royal Institute of Technology
Stockholm, Sweden

ISSN 0302-9743   ISSN 1611-3349 (electronic)
Lecture Notes in Computer Science
ISBN 978-3-031-09826-0   ISBN 978-3-031-09827-7 (eBook)
https://doi.org/10.1007/978-3-031-09827-7

This Springer imprint is published by the registered company Springer Nature Switzerland AG
The registered company address is: Gewerbestrasse 11, 6330 Cham, Switzerland

Preface

This book constitutes the refereed proceedings of the 16th International Conference on Tests and Proofs, TAP 2022, held as part of the federated conference Software Technologies: Applications and Foundations, STAF 2022, Nantes, France, in July 2022.

Six regular papers and one invited extended abstract are included in this volume. The accepted regular papers were selected from eight regular and three short paper submissions following a single-blind review process conducted by an international review committee. The papers are organized into two topical sections: (i) formal analysis and proofs, and (ii) effective testing.

The TAP conference promotes research in verification and formal methods that targets the interplay of proofs and testing: the advancement of techniques of each kind and their combination, with the ultimate goal of improving software and system dependability. Addressing this mission of TAP, at TAP 2022 we had the honour of having Marie-Christine Jakobs (TU Darmstadt) as our keynote speaker, discussing new advancements and challenges in the development of the CoVeriTest framework.

We thank authors of our proceedings papers for revising and submitting their excellent works to the TAP 2022 proceedings. We also thank EasyChair for assisting us in the scientific process of organizing TAP 2022. Finally, we would like to thank the organizing committee of STAF 2022 for their support in making TAP 2022 a successful event.

July 2022

Laura Kovács
Karl Meinke

Organization

Program Committee Chairs

Laura Kovács   TU Wien, Austria
Karl Meinke   KTH Royal Institute of Technology, Sweden

Steering Committee

Bernhardt Aichernig   TU Graz, Austria
Jasmin Blanchette   Vrije Universiteit Amsterdam, The Netherlands
Achim D. Brucker   University of Sheffield, UK
Catherine Dubois (Chair)   ENSIIE, France
Martin Gogolla   University of Bremen, Germany
Nikolai Kosmatov   CEA, France
Burkhart Wolff   University of Paris-Saclay, France

Program Committee

Wolfgang Ahrendt   Chalmers University of Technology, Sweden
Catherine Dubois   ENSIIE, France
Carlo A. Furia   Università della Svizzera italiana, Switzerland
Dilian Gurov   KTH Royal Institute of Technology, Sweden
Falk Howar   Dortmund University of Technology, Germany
Marieke Huisman   University of Twente, The Netherlands
Reiner Hähnle   Technical University of Darmstadt, Germany
Einar Broch Johnsen   University of Oslo, Norway
Konstantin Korovin   The University of Manchester, UK
Jakob Nordstrom   University of Copenhagen, Denmark, and Lund University, Sweden
Patrizio Pelliccione   Gran Sasso Science Institute, Italy
Luigia Petre   Åbo Akademi University, Finland
Cristina Seceleanu   Mälardalen University, Sweden
Sahar Tahvili   Ericsson AB, Sweden
Neil Walkinshaw   The University of Sheffield, UK
Heike Wehrheim   University of Oldenburg, Germany

Additional Reviewers

Jesper Amilon
Richard Bubel
Rong Gu
Jan Haltermann
Marie-Christine Jakobs
Eduard Kamburjan
Christian Lidström
Dominic Steinhöfel
Richard Stewing

Contents

Invited Talk

Automatic Test-Case Generation
with CoVeriTest (Extended Abstract)

Marie-Christine Jakobs[✉][iD]

Technical University of Darmstadt, Department of Computer Science, Darmstadt,
Germany
jakobs@cs.tu-darmstadt.de

Abstract. Automatic test-case generation approaches are applied to
avoid expensive, manual test-case generation. However, it is well-known
that test-case generation approaches come with different strengths and
weaknesses. Therefore, we need to combine different test-case generation
approaches to get stronger test-case generators. The test-case generator CoVeriTest is a framework to combine different test-case approaches.
Its focus is on cooperative, cyclic combinations of test-case generators
based on verification technology. We illustrate how to configure cyclic
combinations in CoVeriTest and explain CoVeriTest's workflow. In more
detail, we discuss the cooperation of the combined test-case generators
and how to distribute the available time among the configured test-case
generators. Thereafter, we give a short overview of the CoVeriTest implementation and report on our observations from our extensive evaluation
of CoVeriTest and on CoVeriTest's performance in the International
Competition on Software Testing (Test-Comp).

1 The CoVeriTest Framework

CoVeriTest [4,5], which stands for **co**operative **veri**fier-based testing, is a
configurable test-case generation approach that generates test cases using verifiers, more specifically reachability analyses like e.g., abstraction-based model
checkers. Its goal is to overcome weaknesses of individual analyses by combining
the strengths of different analyses and supporting cooperation, i.e., information
exchange, between analyses. To avoid covering the same test goal several times
with different analyses, but still continue an analysis after a test goal, which was
difficult to handle for the analysis, was covered by another analysis, CoVeriTest
uses *cyclic* instead of e.g., parallel or sequential combinations of analyses.

To define a cyclic combination in CoVeriTest, the user determines a
sequence of analyses that is executed in every cycle of CoVeriTest and specifies
how analyses cooperate within a cycle or across cycles (see Sect. 1.1). In addition, the user configures the total time limit for CoVeriTest and an (initial)

This work was funded by the Hessian LOEWE initiative within the Software-Factory
4.0 project.

L. Kovács and K. Meinke (Eds.): TAP 2022, LNCS 13361, pp. 3–8, 2022.
https://doi.org/10.1007/978-3-031-09827-7_1

individual time limit for each analysis that will be used to determine how long the analysis may run in a particular cycle.

Given the above information on the CoVeriTest configuration, a specification of the test goals (e.g., one of the Test-Comp specifications [1]), and the program under test, CoVeriTest then cyclically executes the given sequence of analyses until the total time limit is reached or all reachable test goals are covered. In each cycle, the analyses are executed in sequence and each analysis is limited to the minimum of (a) the time limit computed for it for the current cycle and (b) the remaining total time. At the end of each cycle, CoVeriTest determines the time limit of each analysis for the upcoming cycle (see Sect. 1.2).

Before executing an analysis in the current cycle, CoVeriTest considers the configured cooperation and the information provided by previous executions of the same or different analyses to set up the analysis' initial information. In addition, CoVeriTest informs the analysis about the open (not yet covered) test goals. During its execution, the analysis aims to generate tests from counterexamples [2]. To this end, it tries to verify the property that none of the open test goals is reachable. Whenever the analysis detects a property violation, it generates a feasible counterexample that describes an executions that violates the property, i.e., that reaches at least one test goal. Following the approach of Blast [2], the counterexample is then transformed into a test (vector), i.e., a sequence of test inputs for parameters and external functions, which may e.g., be described in the Test-Comp format [1]. After removing the goals covered by the generated test from the set of uncovered goals and updating the explored state space accordingly, the analysis continues until all goals are covered, all open test goals are proven unreachable, or the analysis' time limit exceeds. At the end of the analysis, the analysis provides the generated test cases, the covered goals, and information about its performed analysis, e.g., the explored state space.

1.1 Cooperation

The goal of cooperation in CoVeriTest is to let analyses profit from knowledge gathered by previous executions of analyses about infeasible paths, already explored paths, etc. To enable cooperation, after each execution an analysis provides information about its explored state space in form of an abstract reachability graph (ARG). Before executing an analysis, CoVeriTest considers those ARGs to compute the analysis' initial information. The computed information depends on the configured cooperation, which can be chosen per analysis. Depending on the analysis, CoVeriTest currently supports up to nine different cooperation choices. Next to (a) no cooperation (i.e., the ARGs are ignored and the analysis starts from scratch), CoVeriTest allows (b) cooperation between different executions of the same analysis, (c) between different analyses and (d) certain combinations of (b) and (c).

For *cooperation between different executions of the same analysis*, we always need to exchange information across different cycles. Since the cooperation setting of an analysis is fixed, we assume that an analysis' most recent ARG, i.e., its ARG from the previous cycle contains at least as much information as all other previous ARGs of the analysis. Therefore, we only consider the analysis' ARG of the previous cycle when computing the initial information. So far, CoVeri-Test supports two choices to setup the initial information: either (b1) reusing the precision (i.e., the degree of abstraction) used in the previous execution or (b2) reusing the ARG (including the precision). In the former case, we restart the analysis with the precision from the previous execution, which might allow us to avoid abstraction refinements for spurious counterexamples once found by the analysis. In the latter case, we continue the exploration of the previous analysis.

Cooperation between different analyses may happen within or across cycles. While one may use any previously provided ARG, more recent ARGs likely provide more useful information. For instance, the analysis producing them might already reuse information and may be more up-to-date with respect to the covered and open test goals. In favor of limiting the required storage as well as the configuration space, CoVeriTest only considers the most recent ARG, i.e., the ARG produced by the last analysis execution of all analyses executions so far, when cooperating between analyses. Again, CoVeriTest provides two options to profit from the most recent ARG. Option (c1) applies the idea of conditional model checking [3]. It uses the ARG to compute a condition describing the non-explored program paths and lets the upcoming analysis use the condition to restrict its exploration to those non-explored paths. Option (c2) allows the analysis to compute an initial precision from the precision used in the lastly executed analysis. To this end, a transformation between the different types of precisions must be supported. Currently, only transformations from predicate to value precisions exists (see e.g., [12]). In contrast to case (b), the two options can also be combined. Furthermore, options (c1) and (c2) can also be combined with option (b1) individually or jointly. Due to technical reasons a combination with (b2), which would require a transformation of the reused ARG from (b2), is not supported.

1.2   Adjusting Individual Analysis Time Limits

Initially, CoVeriTest has used a *fixed time limit* per analysis, namely the analysis' initial time limit. Hence, the time limit of analysis i stays the same in each cycle, i.e., $\text{limit}_i^{\text{new}} = \text{limit}_i^{\text{init}}$. Nowadays, CoVeriTest also supports to *dynamically adjust* the time limit. In case of dynamic adjustment, CoVeriTest uses the initial time limits in the first cycle. At the end of each cycle, it then redistributes the time of a CoVeriTest cycle ($\sum_{j=1}^{n} \text{limit}_j^{\text{init}}$) among the n analyses of the configured sequence. Since we want to never completely disable an analysis, our redistribution keeps at least a minimum time limit for each analysis and redistributes the remaining time based on the past or predicted future performance of an analysis relative to all other analysis. As *minimum time limit* $t_{\min}$, we choose the minimum of (a) 10 s and (b) the smallest initial time limit ($\min_{i=1}^{n} \text{limit}_i^{\text{init}}$).

When redistributing the time based on the *past behavior*, we look at the number of test goals p_i an analysis newly covered in the last cycle. If no analysis made progress ($p_i \leq 0$ for all analyses $i \in \{1, \ldots, n\}$), we will skip adjusting time limits ($\text{limit}_i^{\text{new}} = \text{limit}_i$). Otherwise we relate the number of test goals newly covered by an analysis to the number of newly covered test goals in the cycle. Since an analysis may cover more goals simply because it had more time, we use a normalization that puts the number p_i in relation to the time limit_i granted to the analysis i. Hence, we update the time limits as follows.

$$\text{limit}_i^{\text{new}} = t_{\min} + \frac{\frac{p_i}{\text{limit}_i}}{\sum_{j=1}^n \frac{p_j}{\text{limit}_j}} * (\sum_{j=1}^n (\text{limit}_j^{\text{init}} - t_{\min})) \qquad (1)$$

While we typically adjust the time limits based on past behavior, for Test-Comp'21 [13] we tried out to distribute the time based on the *predicted future performance* of an analysis. More concretely, we want to grant an analysis more time if we expect the analysis to handle more paths ρ well that cover open test goals. To this end, at the end of each cycle we select a subset of all syntactical paths that cover open test goals and estimate for each path ρ the probability $P(V_i \mid \rho)$ that analysis i detects a feasible counterexample ρ. Since the probability distribution $P(V_i \mid \rho)$ is unknown, we tried to learn it from counterexamples reported by the analysis in separate experiments. Given the learnt distribution, we then adjust the time limits as follows.

$$\text{limit}_i^{\text{new}} = t_{\min} + \mathbb{E}_{\rho \in T}[P(V_i \mid \rho)] * (\sum_{j=1}^n (\text{limit}_j^{\text{init}} - t_{\min})) \qquad (2)$$

2 CoVeriTest Implementation and Evaluation

CoVeriTest is implemented in the software-analysis framework CPA-checker [6]. CPAchecker is highly configurable, already supports conditional model checking, and provides several reachability analyses for C programs including explicit model checking [9], predicate model checking [7], bounded model checking, and symbolic execution [8], which we used for our evaluation. To integrate CoVeriTest into CPAchecker, we basically implemented two algorithms: (a) a circular algorithm to configure and cyclically run the sequence of analyses as well as to realize the information exchange (cooperation) and (b) a test-case generation algorithm, which wraps a reachability analysis to intercept found counterexamples, transform them into tests, and then continue the analysis. To provide the test-goal specification to the analyses, we enrich the analyses with an observer component monitoring the open, but not yet covered test goals. Currently, our implementation only considers test-goals that are a subset of a program's control-flow edges like branches and optionally performs initial test goal reduction [10,14]. In principle, CoVeriTest works with any test goal automata operating on syntactical program elements and, thus, e.g., arbitrary test-goal specifications in FQL [11].

To gain insights how to configure CoVeriTest, we performed an extensive experimental comparison of 126 CoVeriTest configurations [4,5] resulting from combining two sequences of analyses (value analysis plus predicate analysis and bounded model checking plus symbolic execution) with 6 initial time limits and 12 (9) cooperation settings. We observed that how to configure CoVeriTest to get a well performing CoVeriTest instance depends on the used sequence of analyses. Nevertheless, when choosing a good configuration of time limits and cooperating setting for a given sequence, CoVeriTest's performance (covered test goals) is typically better than using a single analysis of the analyses in the CoVeriTest sequence or their parallel combination.

To compare CoVeriTest against other state-of-the-art test-case generation tools, CoVeriTest has been participating in the annual International Competition on Software Testing[1] (Test-Comp) since the initiation of Test-Comp in 2019. In Test-Comp'19, CoVeriTest uses a combination of explicit and predicate model checking with fixed time limits of 20 s and 80 s per analysis (the best configuration we studied [4,5]). For Test-Comp'20, the CoVeriTest configuration from Test-Comp'19 has been extended with dynamic time limit adjustment based on past behavior (Eq. 1) while in Test-Comp'21 we used dynamic time limit adjustment based on predicted future behavior (Eq. 2). For Test-Comp'22, we submitted a sequential composition of a random test-case generator and the CoVeriTest configuration from Test-Comp'20, i.e., we switched back to time adjustment based on past behavior. Furthermore, we extended CoVeriTest's test export with an additional test-case mutation, which in addition to each test generated by CoVeriTest exports a fixed number of mutations of that test. While not being the best tool, CoVeriTest regularly belongs to the best third of the Test-Comp participants in the category `cover-branches`.

References

1. Beyer, D.: First international competition on software testing. Int. J. Softw. Tools Technol. Transf. **23**(6), 833–846 (2021). https://doi.org/10.1007/s10009-021-00613-3
2. Beyer, D., Chlipala, A., Henzinger, T.A., Jhala, R., Majumdar, R.: Generating tests from counterexamples. In: ICSE, pp. 326–335. IEEE (2004). https://doi.org/10.1109/ICSE.2004.1317455
3. Beyer, D., Henzinger, T.A., Keremoglu, M.E., Wendler, P.: Conditional model checking: A technique to pass information between verifiers. In: Proceedings of FSE, p. 57. ACM (2012). https://doi.org/10.1145/2393596.2393664
4. Beyer, D., Jakobs, M.-C.: CoVeriTest: Cooperative verifier-based testing. In: Hähnle, R., van der Aalst, W. (eds.) FASE 2019. LNCS, vol. 11424, pp. 389–408. Springer, Cham (2019). https://doi.org/10.1007/978-3-030-16722-6_23
5. Beyer, D., Jakobs, M.: Cooperative verifier-based testing with CoVeriTest. Int. J. Softw. Tools Technol. Transf. **23**(3), 313–333 (2021). https://doi.org/10.1007/s10009-020-00587-8
6. Beyer, D., Keremoglu, M.E.: CPAchecker: A tool for configurable Software Verification. In: Gopalakrishnan, G., Qadeer, S. (eds.) CAV 2011. LNCS, vol. 6806, pp. 184–190. Springer, Heidelberg (2011). https://doi.org/10.1007/978-3-642-22110-1_16

---

[1] https://test-comp.sosy-lab.org/.

7. Beyer, D., Keremoglu, M.E., Wendler, P.: Predicate abstraction with adjustable-block encoding. In: FMCAD, pp. 189–197. IEEE (2010). https://ieeexplore.ieee.org/document/5770949/
8. Beyer, D., Lemberger, T.: Symbolic execution with CEGAR. In: Margaria, T., Steffen, B. (eds.) ISoLA 2016. LNCS, vol. 9952, pp. 195–211. Springer, Cham (2016). https://doi.org/10.1007/978-3-319-47166-2_14
9. Beyer, D., Löwe, S.: Explicit-state software model checking based on CEGAR and interpolation. In: Cortellessa, V., Varró, D. (eds.) FASE 2013. LNCS, vol. 7793, pp. 146–162. Springer, Heidelberg (2013). https://doi.org/10.1007/978-3-642-37057-1_11
10. Chusho, T.: Test data selection and quality estimation based on the concept of esssential branches for path testing. IEEE TSE **13**(5), 509–517 (1987). https://doi.org/10.1109/TSE.1987.233196
11. Holzer, A., Schallhart, C., Tautschnig, M., Veith, H.: How did you specify your test suite? In: ASE, pp. 407–416. ACM (2010). https://doi.org/10.1145/1858996.1859084
12. Jakobs, M.C.: Reusing predicate precision in value analysis. In: ter Beek, M.H., Monahan, R. (eds.) IFM 2022. LNCS, vol. 13274, pp 63–85. Springer, Cham (2022). https://doi.org/10.1007/978-3-031-07727-2_5
13. Jakobs, M.-C., Richter, C.: CoVeriTest with adaptive time scheduling (Competition contribution). In: Guerra, E., Stoelinga, M. (eds.) FASE 2021. LNCS, vol. 12649, pp. 358–362. Springer, Cham (2021). https://doi.org/10.1007/978-3-030-71500-7_18
14. Marré, M., Bertolino, A.: Using spanning sets for coverage testing. IEEE TSE **29**(11), 974–984 (2003). https://doi.org/10.1109/TSE.2003.1245299

Formal Analysis and Proofs

REACH on Register Automata via History Independence

Simon Dierl[1]($\boxtimes$) (iD) and Falk Howar[1,2]($\boxtimes$) (iD)

[1] TU Dortmund University, Dortmund, Germany
{simon.dierl,falk.howar}@tu-dortmund.de
[2] Fraunhofer ISST, Dortmund, Germany

Abstract. Register automata are an expressive model of computation using finite memory. Conformance checking of their properties can be reduced to NonEmptiness tests, however, this problem is PSPACE-complete. Existing approaches usually employ symbolic state exploration. This results in state explosion for most complex register automata. We propose a semantics-preserving transformation of register automata into a representation in which reachability of states is equivalent to reachability of locations, i.e., is in NL. We evaluate the algorithm on random-generated and real-world automata and show that it avoids state explosion and performs better on most instances than a comparable existing approach. This yields a practical approach to conformance checking of register automata.

Keywords: Register automata · Non-emptiness · Reachability · History independence · Constraint projection · Conformance checking · Model checking

1 Introduction

Register automata (RAs) were introduced by Kaminski and Frances [24] to model languages over infinite alphabets by combining finite-state automata with a finite set of *registers* that can hold data from the inputs. For example, a register automaton can recognize the language of strings beginning and ending with the same letter from an infinite alphabet by storing the initial symbol in a register and comparing every subsequent input symbol to the register. RAs can be used to model many types of real-world systems. As a result, performing model checking on register automata is a relevant problem. Commonly, model checking is performed via conformance checking: a property is modeled as second RA that accepts if the property is satisfied. Then, the product of both automata is constructed and tested for NonEmptiness. If it is non-empty, the property is satisfied for at least one input in the original automaton. This is equivalent to running the automaton under test and the specification automaton in parallel and checking if both simultaneously accept.

© The Author(s), under exclusive license to Springer Nature Switzerland AG 2022
L. Kovács and K. Meinke (Eds.): TAP 2022, LNCS 13361, pp. 11–30, 2022.
https://doi.org/10.1007/978-3-031-09827-7_2

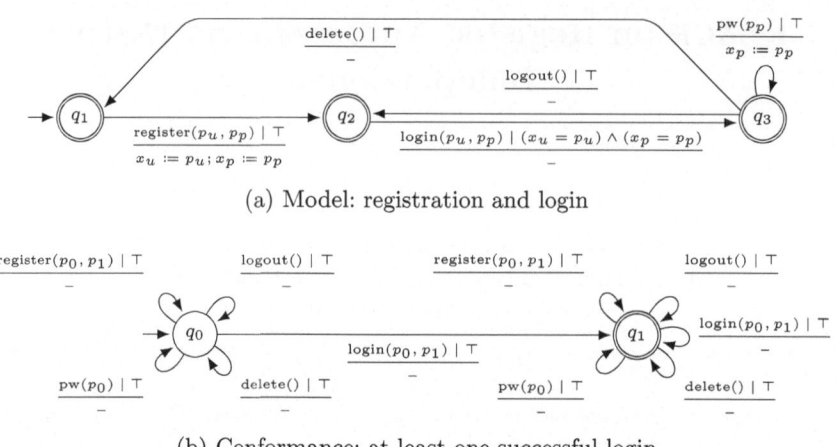

(a) Model: registration and login

(b) Conformance: at least one successful login

Fig. 1. A RA (a) modeling XMPP [35] account management for a single user [9, Fig. 1] and a RA (b) specifying that at least one login can be performed.

As an example, Fig. 1(a) shows an RA recognizing the single-user account management fragment of the extensible messaging and presence protocol (XMPP) [35]. In q_1, no account is registered, in q_2, an account exists, but the user is not logged in, while in q_3, the user is. The user's name and password are stored in registers x_u and x_p, respectively. Figure 1(b) specifies that at least one login must be successful by recognizing inputs that contain at least one login() symbol. The language accepted by the product automaton of (a) and (b) is therefore non-empty if the XMPP fragment accepts logins.

However, NONEMPTINESS (and the more general REACH) are PSPACE-complete for most types of RA (see [16] for a more detailed overview). Two approaches to perform NONEMPTINESS testing are described in the literature. Sakamoto and Ikeda [36] describe a transformation to equivalent finite-state automata that results in exponential blow-up, while other approaches such as D'Antoni et al.'s configuration LRS [14] for symbolic register automata and Boigelot's symbolic state-space exploration [5] perform symbolic forward search from the initial states, which also creates locations exponential in the number of registers.

To improve upon this, we propose a novel approach that transforms a RA into an equivalent *history independent* RA. In these, it is guaranteed that in every reachable state, a matching input for every subsequent transition exists. Therefore, we can reduce REACH for the automaton's execution states to REACH on the transition graph, which is only an NL-complete problem. We achieve this by back-propagating constraints imposed by a subsequent guard and splitting locations when necessary. Instead of tracking every possible register state, this approach only considers properties that are enforced by subsequent guards, greatly reducing the number of symbolic states for many automata. We verify this on both real-world and synthetic automata. Our implementation is publicly available on GitHub.[1]

Related Work. Similar to our approach, Garhewal et al. [18] track location constraints in their SL^* learning algorithm for register automata. They use forward propagation, and discard constraints that do not discriminate accepted suffixes to avoid exponential blow-up. Iosif and Xu [23] also applied incremental refinenemt in a forwards-propagation algorithm for alternating data automata.

Cassel et al. developed RALib [8], another RA library for the JVM. Since we use slightly different RA semantics, we decided to implement our own RA library. Chen et al. [11] also explore emptiness tests for extended theories; similar approaches are used by [7] and [13].

Constraint solving via SMT solvers on Open pNets was used in [25] to perform dead transition removal to enable bisimulation checking and in [33] for single-step satisfiability testing by encoding the complete system as a SMT problem.

Moerman and Sammartino studied residuality [27] in nominal automata [6], a property similar to our history independence property.

2   Preliminaries

This section introduces the subset of first-order logic used by register automata and our proposed algorithm, as well as data words and register automata themselves.

2.1   Logical Operations

First, we define the subset of logical formulae relevant to our problem and operations on them.

Definition 1 (Conjunction of Comparisons, Valuation). *Given a set V of variables, the logic of conjunctions (of comparisons) over V $CC[V]$ is a subset of first-order logic formulae with variables V. It is defined by*

$$C[V] ::= ::=(v_1 \circ v_2) \ for \circ \in \{=, \neq\}, v_1, v_2 \in V$$
$$CC[V] ::= ::=\top \mid \bot \mid C[V] \ (\wedge C[V])^*,$$

where $\top$ is universally true and $\bot$ is unsatisfiable. We refer to elements of $C[V]$ as clauses and v_1, v_2 as variables. The set of free variables is empty if $f = \top$ or $f = \bot$ and otherwise is the set of variables present in one or more clauses.

For a given domain D, a valuation $\phi : V \to D$ satisfies a conjunction $f \in CC[V]$ if and only if $f = \top$ or $f = \wedge_{c \in C} c$ and for all $(v_i \circ v_j) \in C$, $\phi(v_i) \circ \phi(v_j)$. We then write $\phi \models f$.

Two conjunctions f, g are equivalent ($f \equiv g$) if the set of satisfying valuations is identical for both.

---

[1] https://github.com/tudo-aqua/koral.

Our proposed algorithm will make use of two operations on these conjunctions. Given a conjunction of comparisons, a constraint projection generates the strongest statement about a subset of its variables implied by the conjunction. Limiting logical operators to $\wedge$ permits all $CC[X]$ formulae to be represented as a matrix and the following algorithms to be implemented efficiently.

Definition 2 (Constraint Projection). *Let* $f \in CC[V]$ *and let* $W \subseteq V$. *The* (constraint) projection $\Pi_W(f)$ *is a logical formula such that*

- $f \implies \Pi_W(f)$,
- *all free variables in* $\Pi_W(f)$ *are* $\in W$, *and*
- $\forall g$ *with free variables* $\in W$ *and* $f \implies g$, $\Pi_W(f) \implies g$.

For example, $\Pi_{\{x,z\}}(x = y) \wedge (y = z)$ is $(x = z)$, since it is implied by the statements in the original formula. $\Pi_W(\bot) = \bot$ for all W, since it is the strongest possible constraint. In a matrix representation, this operation can be implemented by computing the transitive closure with the Floyd-Warshall algorithm [12] in polynomial time. We next formalize renaming.

Definition 3 (Renaming). *Let* $f \in CC[V]$. *The* renaming $f[v'/v]$ *generates a formula in which all instances of* v *are replaced with* v'. *For vectors* V, V' *of equal size* n, *we write* $f[V'/V] = f[v_1'/v_1, \ldots, v_n'/v_n]$.

2.2   Register Automata

Register automata recognize a combination of a finite and an infinite alphabet. The finite alphabet defines labels that are then combined with values from the infinite alphabet. We now formally define these combinations and mostly follow [9,10,16].

Definition 4 (Data Universe, Symbol, Word). *A* data universe *is a tuple* $\mathcal{D} = (\Lambda, D, \mathfrak{a})$ *with a finite set* Λ *of* labels, *an infinite set* D *of* (data) values, *and an* arity *function* $\mathfrak{a} : \Lambda \to \mathbb{Z}_{\geq 0}$. *For a given label* λ, *the vector of* formal parameters *is* $P^\lambda = (p_1^\lambda, \ldots, p_{\mathfrak{a}(\lambda)}^\lambda)$. *A* data symbol *is a tuple* $(\lambda, \vec{d})$ *with* $\lambda \in \Lambda$ *and a vector of data values* $\vec{d}$ *with* $|\vec{d}| = \mathfrak{a}(\lambda)$. *We usually write a symbol as* $\lambda(d_1, \ldots, d_{\mathfrak{a}(\lambda)})$. *A* data word *is a sequence of data symbols.*

The internal state (q, χ) of a register automaton is defined by its current *location* $q \in Q$ (similar to a finite-state automaton) and the *register valuation* $\chi : X \to D$, i.e., RAs store data values in their registers. Register automata use guard expressions to impose conditions on inputs and their current state. We study guards with equality and inequality comparisons, although in the literature, more expressive guards (e.g., using less-than comparisons) have been discussed. While in the literature, multiple competing formalisms exist, the chosen model subsumes most of them without necessitating expensive transformations (cf. [16]).

Definition 5 (Register Automaton). *A register automaton (RA) is a tuple* $\mathcal{A} = (\mathcal{D}, Q, X, S_0, Q^+, \Gamma)$, *defining*

- *a data universe* $\mathcal{D} = (\Lambda, D, \mathfrak{a})$,
- *a finite set of* locations Q,
- *a finite set of* registers X *that can store data values*,
- *initial states* $S_0 : Q \times (X \to D)$,
- accepting locations $Q^+ \subseteq Q$, *and*
- *a set* Γ *of* transitions $\langle q, q', \lambda, g, u \rangle$, *each defining*
  - *a source location* $q \in Q$,
  - *a target location* $q' \in Q$,
  - *a label* $\lambda \in \Lambda$,
  - *a guard* $g \in \mathsf{CC}[X \cup P^\lambda]$, *and*
  - *an* update $u : X \to (X \cup P^\lambda)$ *that selects new values for the registers visible in the target location, i.e., $u(x) ::= ::=v$ if the value of register or parameter v is copied to x.*

A transition $\langle q, q', \lambda, g, u \rangle$ is always rendered as

$$\frac{\lambda(p_1^\lambda, \ldots, p_{|\mathfrak{a}(\lambda)|}^\lambda) \mid g}{u},$$

where $p_1^\lambda, \ldots, p_{|\mathfrak{a}(\lambda)|}^\lambda$ are the formal parameters, g is the guard and u is a set of parallel updates $x_i ::= ::=v$ with $v \in X \cup P^\lambda$. If no explicit update to a register x_i is given, the update $x_i ::= ::=x_i$ is implicitly assumed. Next, we will define the execution and acceptance semantics of register automata.

Definition 6 (State Transition). *For a register automaton with a transition* $\gamma = \langle q, q', \lambda, g, u \rangle \in \Gamma$, *a state transition is a tuple* $\mathcal{T} = \langle s, s', \gamma, \lambda(d_1, \ldots, d_{\mathfrak{a}(\lambda)}) \rangle$, *defining*

- *a source state* $s = (q, \chi)$,
- *a target state* $s' = (q', \chi')$,
- *an* underlying transition $\gamma = \langle q, q', \lambda, g, u \rangle$, *and*
- *a data symbol* $\lambda(d_1, \ldots, d_{\mathfrak{a}(\lambda)})$ *from* $\mathcal{D}$,

such that g is satisfied under the valuation $\nu : X \cup P^\lambda \to D$ defined as

$$\nu(v) ::= ::= \begin{cases} \chi(v) & \text{if } v \in X \\ d_i & \text{if } v = p_i, \end{cases}$$

and the target valuation χ' is defined by $\chi'(x) = \nu(u(x))$.

Definition 7 (State Transition Sequence). *Given a register automaton A, a state transition sequence (STS) is a sequence of state transitions $\mathcal{T}_1, \ldots, \mathcal{T}_k$ such that for $1 \leq i < k$, the target state of $\mathcal{T}_i$ is the source state of $\mathcal{T}_{i+1}$.*

The sequence is induced by the data word formed by the data symbols of each state transition. If the initial state of $\mathcal{T}_1$ is $\in S_0$, the sequence is initial. If the target state of $\mathcal{T}_k$ is (q_k, χ_k) and $q_k \in Q^+$, the sequence is accepting.

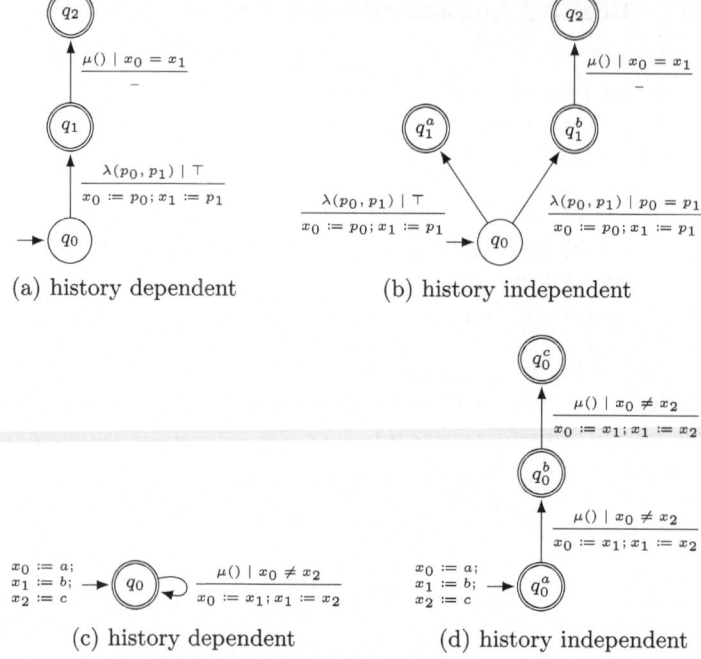

Fig. 2. History dependent and independent RAs accepting the same languages. (a) and (b) accept $\lambda(d, d)\mu() + \lambda(d, e)$; (c) and (d) accept $\mu()^{\{0\text{-}2\}}$.

Definition 8 (Acceptance Behavior). *A register automaton A accepts or rejects* data words *from its data universe. A data word $\lambda_1(\vec{d_1}) \ldots \lambda_k(\vec{d_k})$ is accepted if it induces an initial and accepting STS. A data word that is not accepted is* rejected. *The language of words accepted by the automaton is $L(A)$.*

A is non-empty if and only if $L(A) \neq \emptyset$. We call the corresponding decision problem NONEMPTINESS.

Theorem 1. NONEMPTINESS *is* PSPACE-*complete.* [15]

3   History Independence

We now introduce history independent register automata. Intuitively, a RA is history independent if in each state reachable from an initial state, for every outgoing transition there exists at least one input that enables that transition. As a result, all locations are reachable and transitions can not be "locked out" by register contents. This greatly simplifies analysis of the RA.

Definition 9 (History Independent). *A transition $\gamma = \langle q, q', \lambda, g, u \rangle$ is history independent if for each initial state transition sequence ending in (q, χ),* *there exists a data symbol $\lambda(d_1, \ldots, d_{\mathbf{a}(\lambda)})$ such that there exists a state transition $\langle(q, \chi), (q', \chi'), \gamma, \lambda(d_1, \ldots, d_{\mathbf{a}(\lambda)})\rangle$ for a valuation χ'. A register automaton in which every location is history independent is also* history independent.

Figure 2 shows examples of history dependent and independent automata. Automaton (a) is not independent, since the state $(q_1, \{x_0 \mapsto a, x_1 \mapsto b\})$ with $a \neq b$ conflicts with the transition guard. In this state, no symbol that continues the state transition sequence along the q_1-q_2-transition may exist. Automaton (b) splits q_1 into a terminal version q_1^a and one, q_1^b, with a stronger guard on the incoming edge. This automaton is independent, since in every state (q_1, χ), $\chi \models x_0 = x_1$ and $\mu()$ continues the sequence. Automaton (c) is dependent since after two iterations, $x_0 = x_2$ must hold and the $\mu()$ transition can no longer be taken. Automaton (d) becomes independent by unrolling the transition. After two steps, not more transitions are available.

History independence simplifies analyses on the automaton, since for every path, a matching input must exist, and reachability in the automaton is equivalent to reachability in its graph structure.

Theorem 2. *A history independent automaton is non-empty if and only if one of its accepting locations can be reached from one of its initial locations in the automaton's transition graph structure.*

Proof. If the automaton is non-empty, an accepting input exists. It induces an STS witnessing reachability in the transition graph.

If an accepting location is reachable in the transition graph, a witnessing path $\gamma_1, \ldots, \gamma_k$ exists such that γ_1's source location q_0 has an initial state (q_0, χ_0) Since the automaton is history independent, an input symbol $\lambda(\vec{d})$ must exist that induces a state transition from (q_0, χ_0) to (q_1, χ_1), where q_1 is γ_1's target. Repeat this argument inductively to arrive at (q_k, χ_k), where $q_k \in Q^+$.   □

4   Backwards-Propagation Algorithm

This section introduces our algorithm for transforming register automata into equivalent, history-independent RAs. We begin by introducing location constraints and well-constrained transitions, properties that form a statically verifiable, sufficient condition for history independence. We then outline an algorithm to transform one RA transition into a well-constrained transition. Its iterative application yields a second algorithm that transforms an RA into an equivalent, history independent RA. While the transformation causes an exponential increase in size for some inputs, it is designed to limit the blow-up in most cases. We prove both algorithms' correctness and their worst-case complexity.

4.1   Well-Constrained Transitions

Definition 10 (Location Constraint). *Given a register automaton with location set Q, location constraints are a function $C_Q : Q \to \mathsf{CC}[X]$. If for each initial state (q, χ) and for each initial STS ending in (q, χ), $\chi \models C_Q(q)$, then C_Q is valid in q.*

The constant $\top$ function $q \mapsto \top$ is a valid location constraint for every RA. We also define a shorthand notation for transforming updates into logical constraints.

Definition 11 (Update Constraint) *Given an update* $u = x_1 := v_1, \ldots, x_n := v_n$ *with* $v_i \in X \cup P^\lambda$ *for a label* λ, *the* update constraint *of a is* $C_U(u) = \wedge_{x:=v \in a}(x' = v)$.

We can now construct a property of location constraints and transition guards that is sufficient for proving history independence. This enables us to statically verify history independence instead of reasoning over all inputs.

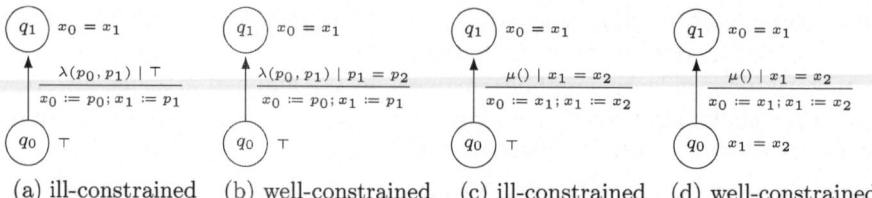

(a) ill-constrained   (b) well-constrained   (c) ill-constrained   (d) well-constrained

Fig. 3. Not well-constrained and well-constrained transitions. Constraints are shown next to the locations.

Definition 12 (Well-Constrained Transition). *Given location constraints* $C_Q : Q \to \mathsf{CC}[X]$, *a transition* $\langle q, q', \lambda, g, u \rangle$ *is* well-constrained *if*

$$C_Q(q) \implies \Pi_X(g), \text{ and}$$
$$C_Q(q) \wedge g \wedge C_U(u) \implies C_Q(q')[X'/X].$$

Well-constrained transitions are essentially equivalent to Hoare triples [20] of form $\{C_Q(q)\}\gamma\{C_Q(q')\}$. Figure 3 shows two examples of non-well-constrained and well-constrained transitions. (a) is not well-constrained, since the guard does not imply the destination constraint. (b) fixes this by strengthening the guard. Note that this may change the semantics of the transition. In (c), the guard is sufficient, but not implied by the source constraint. This is fixed in (d).

Theorem 3. *Given a register automaton with location set* Q *and location constraints* $C_Q : Q \to \mathsf{CC}[X]$, *a well-constrained transition* $\langle q, q', \lambda, g, a \rangle$ *is history independent if* C_Q *is valid in* q.

Proof. Assume a transition $\gamma = \langle q, q', \lambda, g, a \rangle$, an initial STS ending in (q, χ), and location constraints C_Q valid in q such that γ is well-constrained. $g \not\equiv \bot$, since otherwise $\Pi_X(g) \equiv \bot$ and $C_Q(q) \equiv \bot$, i.e., χ could not exist. Now, we show that a valuation $\nu : (X \cup P^\lambda) \to D$ exists that satisfies g and contains χ (i.e., only the parameters could be set). Consider the formula

$$\ell = \bigwedge_{\substack{x,x' \in X \\ \circ \in \{=, \neq\}}} \{(x \circ x') \mid \chi(x) \circ \chi(x')\},$$

which is satisfied by χ. If ν does not exist, $\ell \wedge g$ must be unsatisfiable. Since $g \not\equiv \bot$, the contradiction must pertain to the registers X, i.e., $g \implies (x_i = x_j)$ and $\ell \implies (x_i \neq x_j)$ or vice versa. Then, however, $\Pi_X(g)$ contains $(x_i = x_j)$ and $C_Q(q) \implies g$, so χ can not satisfy $C_Q(q)$. Therefore, a ν extending χ must exist and can be used to select parameter values for P^λ for a state transition. $\square$

Well-constrained transitions also ensure a propagation of valid location constraints.

Theorem 4. *Given location constraints $C_Q : Q \to \mathsf{CC}[X]$ and a state transition $\langle (q, \chi), (q', \chi'), \gamma, \lambda(\vec{d}) \rangle$ with γ being a well-constrained transition, then, if χ satisfies $C_Q(q)$, χ' satsifies $C_Q(q')$.*

Proof. By contradiction. Assume an initial STS ending in $\langle (q, \chi), (q', \chi'), \gamma = \langle y, y', \lambda, g, u \rangle, \lambda(\vec{d}) \rangle$, and location constraints C_Q such that γ is well-constrained, $\chi \models C_Q(q)$ and $\chi' \not\models C_Q(q')$. Then, there must exist x_1, x_2 such that $\chi'(x_1) \neq \chi'(x_2)$, but $C_Q(q') \models x_1 = x_2$ (or vice versa). Now, let $v_1 = a(x_1)$ and $v_2 = a(x_2)$. Since the transition is well-constrained, $g \implies v_1 \neq v_2$, i.e., the transition can not have been taken. $\square$

This informs the central idea of our history-independence transformation: if we can replace all transitions with semantically equivalent well-constrained transitions without altering the automaton's semantics, the resulting automaton is history independent. We will first demonstrate how to make a transition well-constrained using two steps: modifying the guard and updating the source location constraint to make the transition well-constrained and splitting locations to handle incompatible constraints. The first step is straightforward: given a transition $\langle q, q', \lambda, g, u \rangle$ and location constraints C_Q, we can make the transition well-constrained by refining the guard g and location constraint $C_Q(q)$ to

$$g' := g \wedge \Pi_{X \cup P^\lambda}(C_U(u) \wedge C_Q(q'))$$
$$C_Q(q)' := C_Q(q) \wedge \Pi_X(g).$$

If two transitions $\langle q, q'_1, \lambda_1, g_1, a_1 \rangle$ and $\langle q, q'_2, \lambda_2, g_2, a_2 \rangle$ originate in the same location q, no $C_Q(q)$ may exist that makes both transitions well-constrained. In this case, we must split the location into multiple variants, each with the same incoming transitions and loops. To preserve possible determinism, each split would create at least four locations:

- a location q_{12} with $C_Q(q_{12}) = \Pi_X(g_1) \wedge \Pi_X(g_2)$, in which both transitions are present and well-constrained,
- two locations $q_{1\bar{2}}$ and $q_{\bar{1}2}$ with $C_Q(q_{1\bar{2}}) = \Pi_X(g_1) \wedge \Pi_X(\neg g_2)$ and vice versa, in which only one transition is present and well-constrained, and
- a location $q_{\bar{1}\bar{2}}$ with $C_Q(q_{\bar{1}\bar{2}}) = \Pi_X(\neg g_1) \wedge \Pi_X(\neg g_2)$, in which no transition is present.

However, the resulting constraints may not be expressible in $CC[X]$ if they contain $\lor$ operators. Admitting these operators would preclude efficient implementation, so the locations must instead be split into sub-locations for each disjunctive clause. This will frequently result in exponential blow-up. Instead, we sacrifice determinism and create only three locations:

- two locations q_1 and q_2 with $C_Q(q_1) = \Pi_X(g_1)$ and vice versa, in which only one transition is present and well-constrained, and
- a location $q_\top$ with $C_Q(q_\top) = \top$, in which no transition is present.

Note that if a location is split multiple times, $q_\top$ does not need to be recreated. Loop transitions pose an additional challenge, since the location constraint required to make it well-constrained may vary between loops. To preserve semantics, a loop $\langle q, q, \lambda, g, u \rangle$ must be split into two transitions:

- $\langle q_l, q, \lambda, g' = \Pi_{X \cup V}(C_U(u) \land C_Q(q)), u \rangle$ with $C_Q(q_l) = \Pi_X(g')$, which is well-constrained, and
- $\langle q_l, q_l, \lambda, g, u \rangle$, which may not.

However, the new loop can again be split iteratively until a stable state is reached. We will prove this intuition in the next section in Lemma 1. We formalize our idea as Algorithm 1. Figure 4 shows two examples of the algorithm's operation. In (a) and (b), q_1' is created to accommodate the μ transition. The guard and the location constraint of q_1' are strengthened so the transition is now well-constrained. Note that the new q_0-q_1' transition is not well-constrained. In (c) and (d), loop handling is shown. The unrolled iteration (the q_1'-q_1 transition) is well-constrained, the copied loop is not.

Note that the loss of determinism is significant, since deterministic RAs are a strict subset of nondeterministic ones. E.g., the former are closed under complement and their UNIVERSALITY is decidable, while nondeterministic RAs are not closed under complement and UNIVERSALITY is undecidable. This does not impact our REACH analysis, however, which will benefit from the transformation. Next, we prove that the algorithm does not alter an RA's semantics, conditional on the validity of the location constraints.

Theorem 5. *Let A be an RA with location constraints C_Q and A^* the RA with location constraints C_Q^* after applying Algorithm 1 to a single transition $\gamma = \langle q, q, \lambda, g, u \rangle$. A data word induces an accepting initial STS in A such that in each state $(\bar{q}, \bar{\chi})$, $\bar{\chi} \models C_Q(\bar{q})$ if and only if it induces an accepting initial STS in A^* such that $\bar{\chi} \models C_Q^*(\bar{q})$.*

Proof. State transition sequences may include three types of modified transition. For each type, we show that an original transition can be exchanged with the modified transition and vice versa.

Outgoing transitions For STS in A^*, they can be replaced by the original transition with a weaker guard. For STS in A, assume a state transition $\langle (q, \chi), (q', \chi'), \gamma, \lambda(\vec{d}) \rangle$. Since $\chi' \models C_Q(q')$ and the original guard is satisfied, the stronger guard created in Line 4 must also be satisfied, so the transition can be replaced with the modified variant.

```
 1 Function MakeWellConstrained(A, C_Q, γ = ⟨q, q', λ, g, u⟩) is
 2 |   remove γ from A;
 3 |   SplitLocation(A, C_Q, q, ⊤);
 4 |   g* := Π_{X∪V}(a ∧ C_Q(q'));
 5 |   c := Π_X(g');
 6 |   q_c := SplitLocation(A, C_Q, q, c);
 7 |   add transition γ* = ⟨q_c, q', λ, g*, u⟩ to A;
 8 |   if q = q' then
 9 |   |   add transition ⟨q_c, q_c, λ, g, u⟩ to A;
10 |   |   add transition ⟨q_c, q_⊤, λ, g, u⟩ to A;
11 |   return γ*;

12 Function SplitLocation(A, C_Q, q, c) is
13 |   if q_c already exists in A then return q_c;
14 |   add a new location q_c to A;
15 |   C_Q(q_c) := a;
16 |   foreach incoming transition ⟨q̂, q, λ, g, u⟩ do
17 |   |   add transition ⟨q̂, q_c, λ, g, u⟩ to A;
18 |   foreach loop transition ⟨q, q, λ, g, u⟩ do
19 |   |   add transition ⟨q, q_c, λ, g, u⟩ to A;
20 |   if q is accepting then mark q_c as accepting;
21 |   if ∃ initial state (q, χ) with χ ⊨ c then add initial state (q_c, χ);
22 |   return q_c;
```

Algorithm 1: Well-constrained transformation for single transitions.

Loop transitions For STS in A^*, they can again be replaced with the original. For STS in A, the loop may have been split. Then, all but the penultimate loop iteration can be replaced with the loop copy made by Line 9; the last iteration then uses the transition to the original location. If not, but the transition after the loop was split, the last loop iteration can be replaced with the copy made in Line 19.

Incoming transitions For STS in A^*, replace the transition with the original variant. For STS in A, the replacement transition must be selected on the following transitions. If the transition is the last, select the transition to $q_⊤$. If the next transition is $γ$, select the transition to the split-off location, if not, to the original.

Acceptance also remains identical due to Line 20. For initial locations, the argument is similar to outgoing transitions. If the first transition is $γ$, the original guard was satisfied and $χ' ⊨ C_Q(q')$, so the initial valuation $χ ⊨ C_Q^*(q_c)$ and q_c is initial.   □

4.2   Iterative History Independence Transformation

To make an RA history independent, we can repeatedly apply Algorithm 1 until all transitions are history independent. This is formalized in Algorithm 2.

We will show the algorithm's correctness in multiple parts: first, we will demonstrate termination, then argue that the final set of location constraints is valid, and conclude that the resulting automaton must be history independent and its semantics are unchanged.

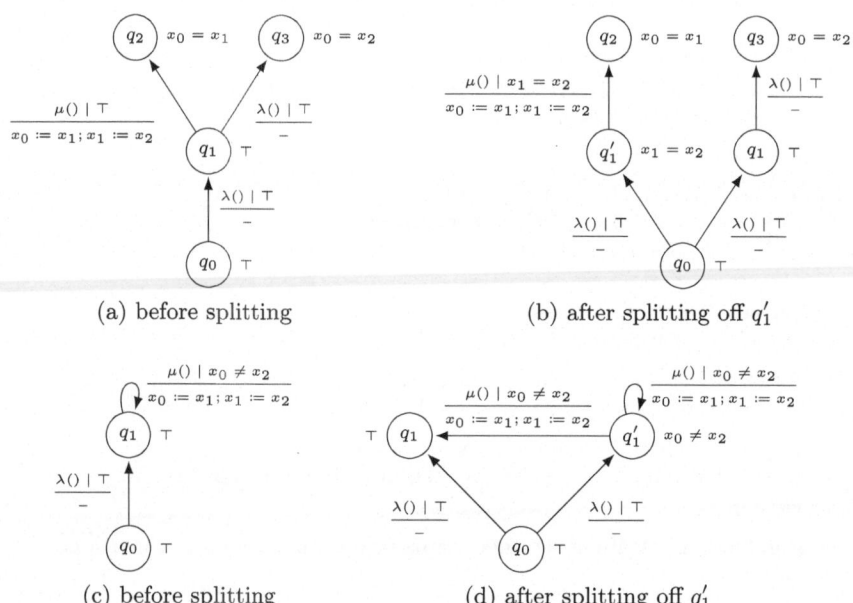

(a) before splitting

(b) after splitting off q'_1

(c) before splitting

(d) after splitting off q'_1

Fig. 4. Example results of the `MakeWellConstrained` function of Algorithm 1. (a) is transformed to (b) and (c) is transformed to (d).

Lemma 1. *Algorithm 2 terminates.*

Proof. The algorithm's state space is described by (Q_c, Γ_H), with Q_c being the constraint-specific locations and Γ_H the well-constrained transitions. We first define a lattice over the locations with their constraints. A location set Q_c is smaller than Q'_c if for every original location, the constraints of Q'_S have become stricter:

$$Q_c \sqsubseteq Q'_c \iff \forall q \in \text{original } Q \forall q_c \in Q_c \exists q'_c \in Q'_c : c \impliedby c'.$$

Note that $CC[X]$ is finite. This allows us to define the state space as a lattice, with

$$(Q_c, \Gamma_H) \sqsubseteq (Q'_c, \Gamma'_H) \iff Q_c \sqsubseteq Q'_c \lor (Q_c = Q'_c \land \Gamma_H \subseteq \Gamma'_H).$$

In each step, the algorithm either marks a transition as well-constrained and leaves the automaton unchanged (Line 13), or splits a location and therefore strengthens its constraint. Therefore, it terminates in a finite number of steps due to the fixed point theorem of Knaster and Tarski [19, §2.1]. □

```
1  Function MakeHistoryIndependent(A) is
2  │   foreach location q do C_Q(q) := ⊤;
3  │   Γ_H := ∅;
4  │   while ∃γ ∉ Γ_H do
5  │   │   γ_H := MakeWellConstrained(A, C_Q, γ);
6  │   └   add γ_H to Γ_H;
7  └   return (A, C_Q);
```

Algorithm 2: History Independence Transformation for RAs

Lemma 2. *The location constraints returned by Algorithm 2 are valid for A.*

Proof. Since the algorithm terminates (Lemma 1), all transitions must be well-constrained transitions. For each initial STS, Line 21 ensures that the initial states satisfy their locations constraints. Inductively, by Theorem 1, if the nth states satisfies its constraint, so will the $n + 1$st. Therefore, the constraints are valid for each reachable state.   □

Theorem 6. *Algorithm 2 computes an history independent automaton accepting the same language as the input RA.*

Proof. Since Lemma 2 guarantees that all locations constraints are valid. All transitions are well-constrained, so by Theorem 3, they must be history independent. Because the initial constraints ($\top$) are valid for the input RA by definition, Theorem 5 guarantees that for each data word inducing an accepting initial STS in the original RA, an accepting initial STS in the history independent RA exists and vice versa, i.e., the accepted languages are identical.   □

Finally, we show that the problem of transforming an RA into an equivalent history independent RA is FPSPACE-complete. This follows from NonEmptiness being PSPACE-complete in general, but simple to check on an history independent automaton.

Corollary 1. *History independence is FPSPACE-complete.*

Proof Sketch. We begin by showing hardness by reducing the NonEmptiness problem on RAs. NonEmptiness for an arbitrary RA is PSPACE-complete, but NonEmptiness of a history independent register automaton can be tested by checking if an accepting location can be reached from an initial state's location in the transition graph. Since Reach on directed graphs in $\in$ NL, the history independence transformation must be at least FPSPACE-hard.

Now, we show that the transformation can be implemented using polynomial space. The state space of Algorithm 2 is bounded by the maximal number of transitions. For each original location, up to $3^{|X|^2}$ split-off versions may exist. The automaton size is at worst $|Q| \cdot 3^{|X|^2}$ (i.e., exponential in the original automaton's size). However, each state can be identified using a string of polynomial length, admitting an FPSPACE implementation.   □

5   Evaluation

We implemented our back-propagating (BP) algorithms in Kotlin in the KORAL software package and compared our performance to a symbolic *forward*-propagating (FP) conformance checker based on RALib[2] [8], JConstraints[3] [21] and Z3[4] [28].

As benchmarks, we used real-world-derived RAs provided in the Automata Wiki[5] [31] as well as random-generated register automata. For the Wiki automata, we simulated a conformance checking (CC) scenario by using the same automata as SUT and specification, i.e. we analyzed the self-product of each automaton.

We analyzed the performance of both our implementation and the RALib-based analyzer in the conformance checking scenario. Additionally, we studied our algorithm's performance on the original Wiki automata and the random RAs. For our algorithm, we recorded both the blowup (BU) in automaton size and the execution time. For the FP analyzer, only timing data was available. The experiments were executed on a Java VM with 32 of heap memory running on an Intel Core i9-7960X CPU. A Docker image of the experimental setup is available on Zenodo [17].

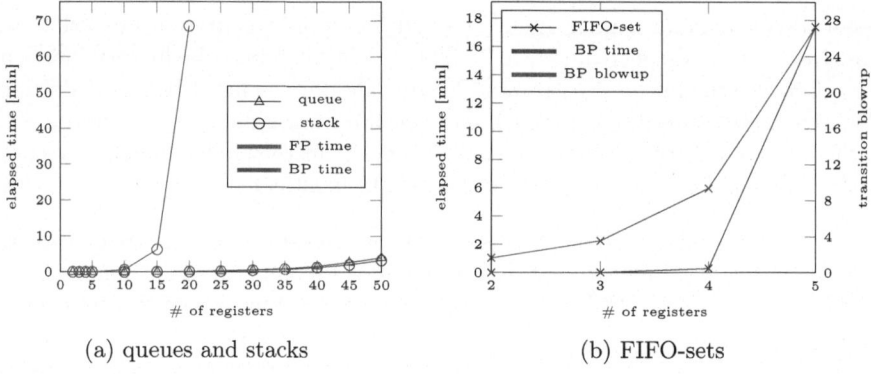

(a) queues and stacks   (b) FIFO-sets

Fig. 5. Evaluation results for data structure conformance checks. For all structures, the time required for successful forward- and back-propagation is shown, for FIFO-sets, the transition blowup caused by back-propagation is also rendered.

5.1   Automata Wiki Benchmarks

The Automaton Wiki provides three sets of data structure benchmarks. These sets model bounded queues, stacks and FIFO-sets (stacks with a uniqueness con-

[2] https://bitbucket.org/learnlib/ralib/src/eqmc.
[3] https://github.com/tudo-aqua/jconstraints.
[4] https://github.com/Z3Prover/z3.
[5] https://automata.cs.ru.nl.

straint). The bound is determined by the number of registers; the wiki provides samples for up to 50 registers. The results of our evaluation are shown in Fig. 5. Forwards-propagation failed to analyze all queue and FIFO-set instances and all stacks with $\geq$ 25 registers. Below that, its performance is substantially worse than that of back-propagation. Since the automata contain no guards, BP only verifies that every transition is well-constrained and terminates without modifying the automaton. FIFO-sets demonstrate the limits of our analysis. These contain complex guards and our analysis times out after two hours for $\geq$ 10 registers and the analysis time increases sharply for five registers. The transition blowup (Γ-BU, i.e., the number of generated transitions divided by the number of original transitions) shows an immense increase in the automaton size caused by the transformation; the state blowup (Q-BU) is proportional and therefore not shown. The performance on non-product instances is comparable.

Table 1. Evaluation results on other real-world-derived automata. For all automata, the size, back-propagation performance on original and self-product (conformance check scenario) and forward-propagation performance on the product are shown.

Automaton	$\|Q\|$	$\|\Gamma\|$	BP single			BP CC			FP CC
			Q-BU	Γ-BU	Time	Q-BU	Γ-BU	Time	Time
ABP [4] Output[a]	30	50	1.00	1.00	0.08 s	0.80	0.92	0.20 s	DNF[c]
ABP [4] Receiver 3[a]	6	13	3.83	2.92	0.09 s	10.83	4.26	1.64 s	DNF[c]
ABP [4] Channel[a]	5	8	2.00	1.88	0.05 s	1.43	1.50	0.20 s	DNF[c]
FWGC[b] [32]	18	43	2.89	1.79	0.17 s	7.82	3.74	2293.08 s	0.84 s
Login [1]	12	20	1.25	1.15	0.04 s	2.50	2.23	0.12 s	0.55 s
Map	17	28	1.35	1.36	0.05 s	0.98	1.00	0.86 s	0.94 s
Overwriting Map	15	24	1.53	1.58	0.04 s	0.77	0.88	0.46 s	0.99 s
Passport[a] [3,22]	35	82	1.29	1.12	0.10 s	1.57	1.22	0.47 s	1.04 s
Repdigit Palindrome	6	23	4.00	1.78	0.04 s	10.33	1.92	0.18 s	0.45 s
SIP [1,2,34]	27	65	1.00	1.00	0.04 s	1.13	0.93	0.11 s	43.42 s

[a] automaton was manually transformed into our semantic
[b] farmer, wolf, goat, and cabbage puzzle
[c] execution failed with an error

The results obtained for the remaining instances in the Automaton Wiki are shown in detail in Table 1. Two families of models that require more expressive semantics (randomness and arithmetic) were discarded. Some models were transformed to match our RA semantic without changing their input languages. While most instances were solved efficiently by both implementations, two edge cases stand out: backwards-propagating analysis is superior for the SIP implementation, while it exhibits extremely slow behavior (`MakeWellConstrained` is called 404120 times) for the farmer, wolf, goat, and cabbage puzzle self-product without causing substantial blowup. The latter is due to the automaton having many registers, only two labels, and many loops, resulting in many location combinations in the product automaton which are not reachable from the initial states. Backwards propagation can only discover this by propagating location constraints along all possible paths instead of working "towards" a goal.

Other points of interest are that RALib failed to load the ABP models due to internal errors. Some models (e.g., APB Output and the map self-product) are history independent (blowup ≤ 1). Our algorithm occasionally produced a reduced automaton by removing dead locations or transitions from the product automata (blowup < 1).

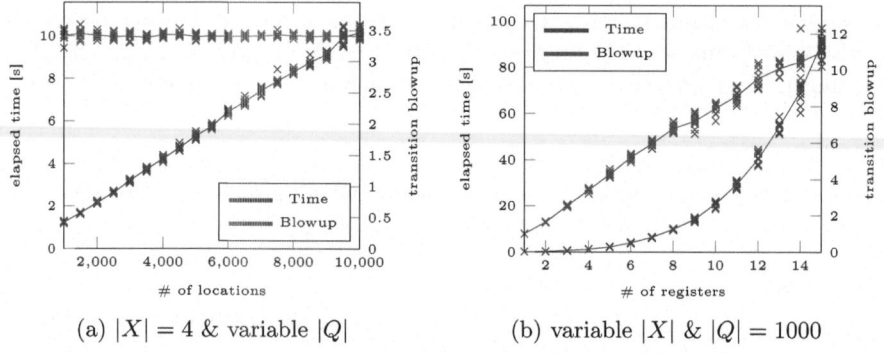

(a) $|X| = 4$ & variable $|Q|$ (b) variable $|X|$ & $|Q| = 1000$

Fig. 6. Back-propagation evaluation results on random RAs. Arity is ≤ 4, 10 automata were generated per parameter set. The line follows the average results.

5.2 Random Automata

To evaluate the impact of several automaton properties, we implemented a random RA generator as part of KORAL and ran our algorithm on families of generated automata. The generator offers three relevant tuneable parameters: maximum arity of symbols, number of registers and number of locations and transitions density. Preliminary analyses showed the impact of the maximum arity to be low (a minor increase in runtime), so we fixed the maximum arity at four. Figure 6 shows the result of the two remaining analyses: increasing number of locations with four registers and increasing number of registers with 1000 locations. All other parameters are fixed. As expected, the runtime is linear in the number of locations with a constant blowup of ca. 3.5. This shows that while more propagation steps are performed, their complexity remains constant. For the number of registers, a polynomial increase in runtime and a roughly linear increase in blowup can be observed. Again, this is to be expected: the back-propagation step is in $\mathcal{O}(|X|^3)$ and more registers enable more possible location constraints.

5.3   Threats to Validity

Since the algorithm is deterministic, all results for the automaton structure are fully replicable. In addition, our timing results were stable for multiple repetitions of the experiments. The performance on random automata is also stable and exhibits low variance. We therefore conclude that the results are internally valid.

However, while the Automaton Wiki automata are representative of real-world problems by design, only ten out of twelve families of automata could be analyzed by our tool. The random automata, while available in arbitrary number, are not necessarily representative of the real world. Therefore, the limited number of real-world automata available for study may limit external validity of our results.

6   Conclusion

Our algorithm provides a novel approach to NONEMPTINESS and REACH checking on register automata. The RA is first transformed into an equivalent, history independent RA. Then, REACH queries, including NONEMPTINESS, can be performed efficiently on the result. The performance of the combined process hinges on our backwards-propagation algorithm.

Compared to previous approaches, this algorithm provides a substantial increase in the number of automata that can be analyzed successfully. We showed that many real-world automata are history independent by design, and for many more, only a small penalty is incurred by the transformation. Our analysis also shows that there exist cases (e.g., the FIFO-set family) in which our approach, while better that prior ones, still performs badly. We additionally found a single case (FWGC self-product) in which our performance is substantially worse than that of previous approaches. Our experiments on random automata demonstrate that our algorithm can scale to very large automata. Its performance degenerates only in the presence of specific structures, not because of size alone.

We conclude that for most "interesting" real-world automata, the analysis speed and blowup do not reach exponential level, since their guard statements rarely need to be traced back multiple steps through a register automaton. Our approach is therefore suitable to analyze a larger subset of real-world register automata than previous approaches.

Future Work. Two independent directions of extension for our work are possible: extending the approach to more expressive languages and adapting the technique to more natural model checking approaches.

Our work can be naturally extended to languages over ordered fields such as $\mathbb{Q}$ and $\mathbb{R}$ with guards containing inequality operators. Recognizing languages over non-field structures is a more challenging task, since e.g. $(x < y) \wedge (y < z)$ can not be statically checked for satisfiability (consider $x = 1, z = 2$). Corresponding models were studied in, e.g., [7,11]. Finally, an approach for register automata with a stack [29,30] may be possible, but would require a backwards-propagation-compatible symbolic representation of stack states. When adding arithmetic operations to register automata, the resulting model becomes a Turing-complete counter automaton [26]. By adapting techniques such as bounded model checking, transformation of some of these automata into a history independent representation would still be possible.

Defining test automata to perform model checking is cumbersome and creating product automata results in a doubling of the number of registers and a corresponding performance penalty. We intend to explore the use of our approach to assist in the efficient evaluation of queries in logical languages such as LTL to offer a more natural and possibly performant approach to model checking.

References

1. Aarts, F., Heidarian, F., Kuppens, H., Olsen, P., Vaandrager, F.: Automata learning through counterexample guided abstraction refinement. In: Giannakopoulou, D., Méry, D. (eds.) FM 2012. LNCS, vol. 7436, pp. 10–27. Springer, Heidelberg (2012). https://doi.org/10.1007/978-3-642-32759-9_4
2. Aarts, F., Jonsson, B., Uijen, J.: Generating models of infinite-state communication protocols using regular inference with abstraction. In: Petrenko, A., Simão, A., Maldonado, J.C. (eds.) ICTSS 2010. LNCS, vol. 6435, pp. 188–204. Springer, Heidelberg (2010). https://doi.org/10.1007/978-3-642-16573-3_14
3. Aarts, F., Schmaltz, J., Vaandrager, F.: Inference and abstraction of the biometric passport. In: Margaria, T., Steffen, B. (eds.) ISoLA 2010. LNCS, vol. 6415, pp. 673–686. Springer, Heidelberg (2010). https://doi.org/10.1007/978-3-642-16558-0_54
4. Bartlett, K.A., Scantlebury, R.A., Wilkinson, P.T.: A note on reliable full-duplex transmission over half-duplex links. Commun. ACM **12**(5), 260–261 (1969). https://doi.org/10.1145/362946.362970
5. Boigelot, B.: Symbolic methods for exploring infinite state spaces. Ph.D. thesis, Université de Liège, Liège, Belgium, May 1998. https://hdl.handle.net/2268/74874
6. Bojańczyk, M., Klin, B., Lasota, S.: Automata theory in nominal sets. Log. Methods Comput. Sci. **10**(4), 1–44 (2014). https://doi.org/10.2168/LMCS-10(3:4)2014
7. Brütsch, B., Landwehr, P., Thomas, W.: N-memory automata over the alphabet N. In: Drewes, F., Martín-Vide, C., Truthe, B. (eds.) LATA 2017. LNCS, vol. 10168, pp. 91–102. Springer, Cham (2017). https://doi.org/10.1007/978-3-319-53733-7_6
8. Cassel, S., Howar, F., Jonsson, B.: RALib: a LearnLib extension for inferring EFSMs. In: Proceedings of the 4th International Workshop on Design and Implementation of Formal Tools and Systems (2015). https://www.faculty.ece.vt.edu/chaowang/difts2015/papers/paper_5.pdf
9. Cassel, S., Howar, F., Jonsson, B., Merten, M., Steffen, B.: A succinct canonical register automaton model. In: Bultan, T., Hsiung, P.-A. (eds.) ATVA 2011. LNCS, vol. 6996, pp. 366–380. Springer, Heidelberg (2011). https://doi.org/10.1007/978-3-642-24372-1_26
10. Cassel, S., Jonsson, B., Howar, F., Steffen, B.: A succinct canonical register automaton model for data domains with binary relations. In: Chakraborty, S., Mukund, M. (eds.) ATVA 2012. LNCS, pp. 57–71. Springer, Heidelberg (2012). https://doi.org/10.1007/978-3-642-33386-6_6

11. Chen, Y.F., Lengál, O., Tan, T., Wu, Z.: Register automata with linear arithmetic. In: 2017 32nd Annual ACM/IEEE Symposium on Logic in Computer Science (LICS), pp. 1–12. IEEE, June 2017. https://doi.org/10.1109/LICS.2017.8005111

12. Cormen, T.H., Leiserson, C.E., Rivest, R.L., Stein, C.: All-Pairs Shortest Paths, chap. 25, 3rd edn. pp. 684–707. MIT Press, Cambridge, February 2009

13. Czyba, C., Spinrath, C., Thomas, W.: Finite automata over infinite alphabets: two models with transitions for local change. In: Potapov, I. (ed.) DLT 2015. LNCS, vol. 9168, pp. 203–214. Springer, Cham (2015). https://doi.org/10.1007/978-3-319-21500-6_16

14. D'Antoni, L., Ferreira, T., Sammartino, M., Silva, A.: Symbolic register automata. In: Dillig, I., Tasiran, S. (eds.) CAV 2019. LNCS, vol. 11561, pp. 3–21. Springer, Cham (2019). https://doi.org/10.1007/978-3-030-25540-4_1

15. Demri, S., Lazić, R.: LTL with the freeze quantifier and register automata. ACM Trans. Comput. Log. **10**(3), 16:1–16:30 (2009). https://doi.org/10.1145/1507244.1507246

16. Dierl, S., Howar, F,; A taxonomy and reductions for common register automata formalisms. In: Olderog, E.-R., Steffen, B., Yi, W. (eds.) Model Checking, Synthesis, and Learning. LNCS, vol. 13030, pp. 186–218. Springer, Cham (2021). https://doi.org/10.1007/978-3-030-91384-7_10

17. Dierl, S., Howar, F.: Reach on register automata via history independence - replication artifact (2022). https://doi.org/10.5281/zenodo.6367981

18. Garhewal, B., Vaandrager, F., Howar, F., Schrijvers, T., Lenaerts, T., Smits, R.: Grey-box learning of register automata. In: Dongol, B., Troubitsyna, E. (eds.) IFM 2020. LNCS, vol. 12546, pp. 22–40. Springer, Cham (2020). https://doi.org/10.1007/978-3-030-63461-2_2

19. Granas, A., Dugundji, J.: Elementary fixed point theorems, chap. 2, pp. 9–84. Springer, New York (2003). https://doi.org/10.1007/978-0-387-21593-8_2

20. Hoare, C.A.R.: An axiomatic basis for computer programming. Commun. ACM **12**(10), 576–580 (1969). https://doi.org/10.1145/363235.363259

21. Howar, F., Jabbour, F., Mues, M.: JConstraints: a library for working with logic expressions in java. In: Margaria, T., Graf, S., Larsen, K.G. (eds.) Models, Mindsets, Meta: The What, the How, and the Why Not? LNCS, vol. 11200, pp. 310–325. Springer, Cham (2019). https://doi.org/10.1007/978-3-030-22348-9_19

22. International Civil Aviation Organization: Machine readable travel documents. Doc 9303, International Civil Aviation Organization, Montréal, Québec, Canada (2021). https://www.icao.int/publications/pages/publication.aspx?docnum=9303

23. Iosif, R., Xu, X.: Abstraction refinement for emptiness checking of alternating data automata. In: Beyer, D., Huisman, M. (eds.) TACAS 2018. LNCS, vol. 10806, pp. 93–111. Springer, Cham (2018). https://doi.org/10.1007/978-3-319-89963-3_6

24. Kaminski, M., Francez, N.: Finite-memory automata. Theor. Comput. Sci. **134**(2), 329–363 (1994). https://doi.org/10.1016/0304-3975(94)90242-9

25. Madelaine, E., Qin, X., Zhang, M., Bliudze, S.: Using SMT engine to generate symbolic automata. Electron. Commun. EASST **76** (2019). https://doi.org/10.14279/tuj.eceasst.76.1103

26. Minsky, M.L.: Computation: Finite and infinite machines. Prentice-Hall International, London (1972)

27. Moerman, J., Sammartino, M.: Residual nominal automata. In: Konnov, I., Kovács, L. (eds.) 31st International Conference on Concurrency Theory. LIPIcs, vol. 171, pp. 44:1–44:21. Schloss Dagstuhl-Leibniz-Zentrum für Informatik, Dagstuhl, Germany (2020). https://doi.org/10.4230/LIPIcs.CONCUR.2020.44

28. de Moura, L., Bjørner, N.: Z3: an efficient SMT solver. In: Ramakrishnan, C.R., Rehof, J. (eds.) TACAS 2008. LNCS, vol. 4963, pp. 337–340. Springer, Heidelberg (2008). https://doi.org/10.1007/978-3-540-78800-3_24

29. Murawski, A.S., Ramsay, S.J., Tzevelekos, N.: Reachability in pushdown register automata. In: Csuhaj-Varjú, E., Dietzfelbinger, M., Ésik, Z. (eds.) MFCS 2014. LNCS, vol. 8634, pp. 464–473. Springer, Heidelberg (2014). https://doi.org/10.1007/978-3-662-44522-8_39

30. Murawski, A.S., Ramsay, S.J., Tzevelekos, N.: Reachability in pushdown register automata. J. Comput. Syst. Sci **87**, 58–83 (2017). https://doi.org/10.1016/j.jcss.2017.02.008

31. Neider, D., Smetsers, R., Vaandrager, F., Kuppens, H.: Benchmarks for automata learning and conformance testing. In: Margaria, T., Graf, S., Larsen, K.G. (eds.) Models, Mindsets, Meta: The What, the How, and the Why Not? LNCS, vol. 11200, pp. 390–416. Springer, Cham (2019). https://doi.org/10.1007/978-3-030-22348-9_23

32. Pressman, I., Singmaster, D.: The jealous husbands and the missionaries and cannibals. Math. Gaz. **73**(464), 73–81 (1989). https://doi.org/10.2307/3619658

33. Qin, X., Bliudze, S., Madelaine, E., Hou, Z., Deng, Y., Zhang, M.: SMT-based generation of symbolic automata. Acta Inform. **57**(3), 627–656 (2020). https://doi.org/10.1007/s00236-020-00367-6

34. Rosenberg, J., et al.: SIP: Session initiation protocol. RFC 3261, RFC Editor, June 2002. https://doi.org/10.17487/RFC3261

35. Saint-Andre, P.: Extensible messaging and presence protocol (XMPP): Core. RFC 6120, RFC Editor, March 2011. https://doi.org/10.17487/RFC6120

36. Sakamoto, H., Ikeda, D.: Intractability of decision problems for finite-memory automata. Theor. Comput. Sci. **231**(2), 297–308 (2000). https://doi.org/10.1016/S0304-3975(99)00105-X

BDDL: A Type System for Binary Decision Diagrams

Yousra Lembachar[1], Ryan Rusich[2], Iulian Neamtiu[3]([✉]),
and Gianfranco Ciardo[4]

[1] Blockdaemon, Los Angeles, CA, USA
yousra@blockdaemon.com
[2] University of California, Riverside, CA, USA
rusichr@cs.ucr.edu
[3] New Jersey Institute of Technology, Newark, NJ, USA
ineamtiu@njit.edu
[4] Iowa State University, Ames, IA, USA
ciardo@iastate.edu

Abstract. Binary Decision Diagrams (BDDs) are compact data structures used to efficiently store and process boolean functions. BDDs have many uses, from system design to model checking to efficiently storing context information for context-sensitive analysis. The use of BDDs in verification and program analysis has been facilitated by the recent emergence of many open source BDD libraries. The correctness of BDD-based system design and verification hinges upon the correctness of the BDD library implementations, and the correct use of these libraries. Surprisingly, for a technology so prevalent in system design and formal verification, there has been little research effort on formally verifying the correctness of BDD library implementations or their use. For BDD libraries that do perform some correctness checks, these are mostly confined to runtime assertion checking, which slows down BDD operations and might still be unable to reveal errors until deployment. To address these issues and take a step toward provably correct, yet efficient, BDD-handling code, we propose a formal system called BDDL to describe, reason about, and prove the correctness of BDD operations. BDDL extends lambda calculus with support for BDD operations (e.g., creation, manipulation), expressing BDD structural properties (e.g., canonicity, proper ordering), and BDD semantics (e.g., sets, relations). BDDL uses a type system based on refinement types to statically check BDD manipulation. We have proved our system correct using a small-step semantics and standard notions of progress and preservation. BDDL is the first attempt to provide a well-defined syntax and semantics to BDD operations; we show how it could prevent bugs and semantic errors in the implementation and use of three mature DD libraries.

Keywords: Binary Decision Diagrams · Type checking · BDD library · Correctness by construction

© The Author(s), under exclusive license to Springer Nature Switzerland AG 2022
L. Kovács and K. Meinke (Eds.): TAP 2022, LNCS 13361, pp. 31–47, 2022.
https://doi.org/10.1007/978-3-031-09827-7_3

1 Introduction

Formal methods for hardware and software verification have been facilitated by reliable and efficient methods for expressing and checking hardware as well as software behavior. For instance, in digital circuit design, where chips can have billions of transistors, symbolic model checking was made possible primarily by the introduction of canonical and efficient data structures such as BDDs, which often provide a compact representation of very large state spaces. Essentially, BDDs can be used to symbolically represent boolean functions. This *symbolic*, rather than explicit representation of the state space is a main strength of BDDs and decision diagrams in general.[1] In addition to symbolic model checking, BDDs are extensively used in quantitative risk assessment; for example, QRAS, a commercial system used by NASA to perform Probabilistic Risk Assessment (PSA) [11], allows systems engineers to quantify risks, identify risk scenarios, as well as reason about how risk is affected by changes to the system or organization— failure for NASA operations can have unacceptable costs.

Numerous decision diagram library implementations support BDDs [1,7,9, 12]. Yet, formal method support for checking BDD correctness is lacking. The aim of this paper is to provide a formal system that verifies the validity of BDD construction and manipulation. The core of our approach consists of a calculus and type system that support BDD terms, BDD operations and BDD semantics. Our current system performs type safety checks for BDD manipulation, but is general enough that we envision it can be extended to support other kinds of decision diagrams. We analyzed three mature DD libraries to drive the design of BDDL. CUDD [7] is a popular DD library, used in the NuSMV model checker. MDDL is part of SMART [9], which has been used to verify the NASA runway safety monitor [20]. JavaBDD [12] is used in bddbddb [22], a Datalog-based framework for specifying, and efficiently performing, program analysis.

To illustrate how our system statically prevents semantic errors, we present two examples of BDD library implementation and BDD library usage errors that cause BDD-based programs to crash or silently produce erroneous outcomes. These examples are drawn from CUDD and MDDL; in Sect. 3 we provide the actual code for these, and other, examples. First, consider the BDD::Compose(g,v) operation from the CUDD library; BDD::Compose returns the result of splicing BDD g into the slot currently occupied by variable with index v in the BDD represented by this. Clients can crash the program by passing in an incorrect index v; recent versions of CUDD generate an Unexpected error, while older versions crash with a Segmentation fault. Second, consider the method RelationalProduct(p,r) from MDDL. The method computes the relational product of BDDs p and r, and requires that p have L levels and r have $2 * L$ levels,

---

[1] Other kinds of decision diagrams operate on integers and reals to encode algebraic, arithmetic, and relational functions. Decision diagrams have been employed in areas as diverse as optimization [2], electronic design [24], VLSI CAD [5], Genetics (gene expression analysis [25], data-mining DNA subsequences) [15], NASA safety operations [20], and reliability [23].

as p encodes a set and r encodes a relation. However, library clients can invoke `RelationalProduct` in incorrect ways: first, they can invoke it with two sets or two relations, in which case the library silently returns an incorrect result; or, they can invoke it with a set and a relation where r's number of levels is not twice p's number of levels, which leads to a runtime error (assertion failure). We present a detailed account of these and other errors in Sect. 3.

In general, BDD libraries do not check the higher-level semantics of library implementations and client-supplied data, or perform such checks at runtime; as a result, they silently return an incorrect result, or fail with a runtime error; another disadvantage of runtime checks is that they slow down the execution. In this paper, we make progress toward provably correct and efficient BDD-handling code using BDDL, a calculus we developed. Our approach consists of two main steps. First, BDD library and client code must be expressed in BDDL, e.g., C, C++, or Java code translated to BDDL, and library function types expressed as BDDL types; currently, this approach is manual, though we found the translation to be straightforward, as evidenced by the translations in Sects. 3. Second, the DDL type inference and checking system statically checks the BDDL code and reports typing errors. Section 3 shows how, when using BDDL, we would get a static typing error in the semantic/assertion failure cases we previously discussed—our system prevents certain ill-typed operations on BDDs that may cause programs to crash or produce erroneous outcomes.

Our approach consists of BDDL, a calculus with an associated type system and operational semantics. BDDL's expressiveness and effectiveness derive from its ability to model and verify two kinds of BDD semantic properties: structural and logical. *Structural* semantic properties, e.g., constraints on node and edge manipulation, are captured by our terms and typing system—we model BDD nodes as terms (Sect. 4), and use a refinement-based type system (Sect. 5) to express integrity properties (e.g., the number of levels, set vs. relation encoding, quasi- vs. fully-reduced form). *Logical* semantic properties are captured by refined function types: the BDDL types assigned to library functions permit concise expression of logical properties (pre- and post-conditions), because refinement predicates on our types allow for conjunctions of arithmetic expressions. At the same time, our subtyping system allows polymorphism over BDD structures, which increases expressiveness, especially for generic BDD-manipulating functions. We employ a small-step operational semantics to prove our system correct using standard notions of progress and preservation (Sect. 5.4).

In summary, our work makes two main contributions:

- An exposé of implementation and usage errors in BDD libraries.
- BDDL, a formal system to express and statically verify the safety of BDD library implementations and BDD library uses.

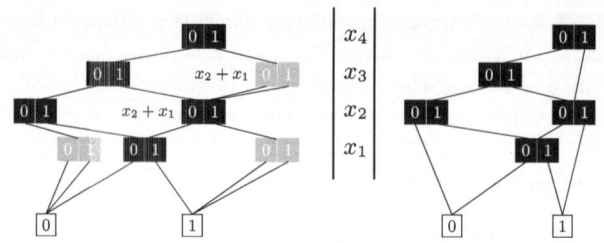

Fig. 1. A quasi-reduced (left) vs. a fully-reduced (right) BDD.

2   Background: Binary Decision Diagrams

We now provide a brief overview of BDDs.

2.1   Binary Decision Diagrams

A Binary Decision Diagram (BDD) is a rooted directed acyclic graph encoding a boolean function of the form $f : \mathbb{B}^L \to \mathbb{B}$, for some $L \in \mathbb{N}^+$. Each non-terminal node is labeled with a variable $x_k \in \{x_L, ..., x_1\}$, and placed at level j (where $j = k$ for BDDs that have the order $x_1 < x_2 < ... < x_L$), while there are two terminal nodes, $\boxed{1}$ and $\boxed{0}$ placed at level 0. A non-terminal node has two outgoing edges labeled "false", or "0", and "true", or "1", respectively, pointing to a node at a level h satisfying $0 \leq h < k$. The value of the function encoded by a BDD for a particular truth value assignment for the L variables is obtained by following the corresponding path from the root to a terminal node. For example, the BDD on the left of Fig. 1 encodes $f(x_4, x_3, x_2, x_1) = \overline{x_4}\,\overline{x_3}x_2x_1 + (x_4 + x_3)(x_2 + x_1)$; note that product denotes logical AND, while sum denotes logical OR. Observe that $f(0, 0, 0, 0) = 0$, since the path from the root corresponding to the choices $x_4 = 0$, $x_3 = 0$, $x_2 = 0$, and $x_1 = 0$ ends at $\boxed{0}$. The BDD on the right encodes the same function, but x_1 is not tested on that path, as it does not affect the value of f when $x_4 = 0$, $x_3 = 0$, and $x_2 = 0$. These two BDDs correspond to quasi-reduced (left) and fully-reduced (right) forms, discussed next; essentially, in fully-reduced BDDs edges can skip levels associated with a variable, whereas in quasi-reduced BDDs edges never skip levels. While the size of a BDD can be exponential in the number of variables, many functions can be encoded compactly.

2.2   Canonical Binary Decision Diagrams

Decision diagrams can be reduced to different forms, so that they still denote the same function in a more compact *canonical* representation. Most library implementations support only canonical forms, having the property that no two nodes encode the same function. This is achieved by first *eliminating duplicate nodes* and then either *retaining* all redundant nodes (quasi-reduced form, no edge skips levels) or *removing* them all (fully-reduced form, edges skip levels whenever possible), where a node is *redundant*, shown in light color in Fig. 1, if its outgoing edges point to the same node.

```
 1 DdNode * Cudd_bddCompose (
 2   DdManager * dd, DdNode * f,
 3   DdNode * g, int v)
 4 {
 5   DdNode *proj, *res;
 6   /* Sanity check. */
 7   if (v < 0 || v >= dd->size)
 8     return(NULL);
 9   proj = dd->vars[v];
10   do {
11     dd->reordered = 0;
12     res = cuddBddComposeRecur(
13            dd,f,g,proj);
14   } while (dd->reordered == 1);
15   return(res);
16 }
```

```
17 BDD BDD::Compose(BDD g,int v)
18 { ...
19   return BDD(...,
20          Cudd_bddCompose(
21          mgr.node, g.node, v));
22 }
23 int main ()
24 {
25   Cudd mgr(0,2);
26   BDD x = mgr.bddVar();
27   BDD y = mgr.bddVar();
28   BDD h = x * y;
29   BDD j = x + y;
30   BDD k = h.Compose(j,2);
31   /* runtime error or crash */
32 }
```

```
 1 letrec two : {ν : nat | ν = 2} =
 2   (succ (succ 0)) in
 3
 4 letrec bddCompose : bdd[l, r, c]
 5   →
      bdd[l', r, c] → {ν : nat | ν ≤ l - 1} =
 6   λ f .
 7     λ g .
 8       λ v .
 9         <body>
10 in
11 letrec h = Bnode(two, ...) in
12 letrec j = Bnode(...) in
13 letrec k =
14   bddCompose h j two // type error
```

C++ code   BDDL code

Fig. 2. Crashing CUDD: the C++ code leads to Segmentation fault in CUDD 2.3.1 and Unexpected error in CUDD 2.4.2.

While other, more specialized, reduction forms have been defined, we only consider the quasi-reduced and fully-reduced forms as they are widely used in decision diagrams libraries. To illustrate why we are interested in these reduction forms, the following section presents some functions from MDDL, the library used in SMART [9], which take BDDs reduced using a specific form as inputs, and we show how BDDL prevents illegal use of such operations.

2.3 Encoding Sets and Relations with BDDs

BDDs can encode sets and relations (we limit the discussion to binary relations, the most widely used in practice). A BDD encodes a set $Y \subseteq \mathbb{B}^L$ through its characteristic function f_Y:

$$i = (i_L, ..., i_1) \in Y \text{ iff } f_Y(i_L, ..., i_1) = 1$$

Analogously, a BDD encodes a relation $R \subseteq \mathbb{B}^L \times \mathbb{B}^L$ through its characteristic function f_R:

$$(i, i') = ((i_L, ..., i_1), (i'_L, ..., i'_1)) \in R \text{ iff } f_R(i_L, i'_L, ..., i_1, i'_1) = 1$$

If the library implementation allows for the same BDD to encode a set or a relation, i.e., a BDD with $2L$ levels may encode a subset of $\mathbb{B}^{2L}$ or a relation over $\mathbb{B}^L$, then the user may use the library in the wrong way: a BDD encoding a set is used as the argument of a function that expects a relation, or vice versa. For example, given an L-level BDD on $(x_L, ..., x_1)$ rooted at p encoding a set $Y \subseteq \mathbb{B}^L$ and a $2L$-level BDD on $(x_L, x'_L..., x_1, x'_1)$ rooted at r encoding a relation $R \subseteq \mathbb{B}^L \times \mathbb{B}^L$, the relational product of p and r returns the root of the L-level BDD encoding the set $\{i' : \exists i \in Y, (i, i') \in R\}$.

In Sect. 3, we consider an implementation of the relational product in MDDL; since root nodes are not checked, the user can pass any ill-typed arguments. We show how BDDL captures these type errors statically.

```
1  BddNode *RelationalProduct (      12  malloc(sizeof(BddNode)); ...
2      BddNode *p, BddNode *r)       13  return bdd;
3  { ...                            14  }
4   ASSERT(( r→ GetLevel()+1)/2      15  int main ()
5          == p→ GetLevel());       16  {
6   ...                             17   BddNode *f=construct_set(2);
7  }                                18   BddNode *g=construct_set(3);
8  BddNode* construct_set (int 1)    19   BddNode *res =
9  {                                20       RelationalProduct(f,g);
10   BddNode *bdd =                  21  /* no exception is raised */
                                     22  }
```

C++ code

```
1  letrec relationalProduct :
2  bdd[l, r, s] → {ν : bdd[l', r, e]
3     | l' = l + l − 1} → bdd[l, r, s] =
4        λ p . λ r . <body>
5  in
6  letrec f : bbd[2, q, s] = Bnode(...)
7  in
8  letrec g : bbd[3, q, s] = Bnode(...)
9  in
10      relationalProduct f g
11      /* type error */
```

BDDL code

Fig. 3. The MDDL relational product code silently outputs an incorrect result when two BDDs encoding a set are passed as arguments.

```
1  BddNode* Union_QQ(               9  }
2      BddNode *p, BddNode *q) {    10 int main ()
3   ASSERT((k = p→ GetLevel())      11 {
4          == q→ GetLevel());       12   BddNode *f = new_bdd(1);
5   ...                             13   BddNode *g = new_bdd(2);
6   ka = answer→ GetLevel();        14   BddNode *res = union(f,g);
7   ASSERT(ka == k);                15   /* runtime error */
8   return answer;                  16 }
```

C++ code

```
1  letrec union : bbd(l, q, c) → bbd[l, q, c]
       → {ν : bbd[l', q, c]) | l' ≤ l} =
2         λ p . λ r . <body>
3  in letrec f : bbd[1, q, s]
4     = Bnode(1, ...) in
5  letrec g : bbd[2, q, s]
6     = Bnode(2, ...) in
7  union f g // type error
```

BDDL code

Fig. 4. The union of two BDDs leads to a runtime error when two BDDs of different levels are passed as arguments (from MDDL).

3   Motivation

The design of BDDL and its type system was driven by examining the source code and evolution (history of bug fixes) of three mature DD libraries: CUDD, MDDL, and JavaBDD. We now proceed to showing the actual code and errors from the examined libraries, the equivalent code in BDDL and the BDDL types that would prevent these errors at compile time.

We begin with the BDD composition code from CUDD (Fig. 2). The left side shows the actual C++ code from the library. Function Cudd_bddCompose (lines 1 through 16) takes BDDs f and g as arguments, and returns the result of splicing g into f at the position indicated by variable with index v. Note the runtime sanity check on line 7, which verifies that the index of v is positive, but less than the BDD size (the number of levels L, as there is one variable per level),

```
1 public class BasicTests extends BDDTestCase {...
2   public void testCrash() {
3     reset(); Assert.assertTrue(hasNext());
4     BDDFactory bdd = nextFactory();
5     BDD a = bdd.one();
6     bdd.reorder(bdd.getReorderMethod());
7 // java.lang.NullPointerException
8 }
```

```
1 letrec reorder :
2 {ν : bbd[l, r, c] | l ≥ 1} → {ν : bbd[l, r, c] | l ≥ 1} =
3   λ p : bbd[l, r, c]. <body>
4 in
5 reorder 1   //type error
```

Fig. 5. JavaBDD: reordering a terminal-only BDD leads to a `nullPointerException`.

i.e., v is within the BDD. The method `BDD:Compose` (lines 17–22) is the C++ library interface for the clients; it invokes `Cudd_bddCompose` on `this.node`, which represents f, and `g.node`, which represents g, and returns a BDD containing the result of the composition. On lines 23–32 we show the actual code of a program we wrote to crash the library. We first construct a BDD h with two levels (x at level 0 and y at level 1, lines 26–28), another BDD j (the number of levels in j is not important in this example), and then invoke `h.Compose(j,2)`, which should splice-in j at index 2. As index 2 does not exist in the BDD, the check on line 7 fails. In the current version, CUDD 2.4.2, this error leads to the program halting with `Unexpected error`; in a previous version we tested, CUDD 2.3.1, it leads to the program crashing with `Segmentation fault`, because the check on line 7 uses `>` rather than `>=`. In either case, the error manifests itself only during execution, and the CUDD client is left with little information as to what the cause of the error is. On the right side of Fig. 2 we show the BDDL code that models the BDD composition via function `bddCompose`; while we have not discussed the BDDL syntax and typing yet, note that our refinement type system allows us to specify that (a) the argument v should be less than the number of levels in f (line 5), and (b) that the argument v has value 2, which is the number of levels in h. Trying to apply `bddCompose` to arguments h j two (line 14) results in a static typing error.

As a second example, we focus on the `RelationalProduct(p,r)` function from MDDL [9], which computes the relational product of BDDs p and r (Fig. 3). As mentioned in Sect. 2.3, the argument p must be an L-level BDD encoding a set, whereas the argument r must be a $2L$-level BDD encoding a relation. MDDL does not differentiate between sets and relations in its actual implementation, therefore, the same BDD may encode a set and a relation and the correct use of these two forms is left to the user.[2] The left side of Fig. 3 contains an excerpt of the library function `RelationalProduct(p,r)` (lines 1–7); the `ASSERT` ensures that the number of levels in r is equal to $2L$ (the +1 is due to levels starting at 0) but no set vs. relation check is performed. The `main` function shows the C++ client code: it uses `construct_set()` to build two BDDs f and g encoding sets (lines 17–18); next, we pass f and g as inputs to `RelationalProduct(p,r)`, which computes the result considering the second argument as a relation. Upon

---

[2] To eliminate ambiguity and prevent a potential incorrect use of functions that support both forms of encoding [1], some libraries do not leave this choice to the user.

completion (line 20), no error is signaled, and the incorrect result is silently returned to the client. On the right side of Fig. 3 we show the BDDL translation. Note how (a) the refinement types of l and l_e on relationalProduct's signature express the runtime checks in the C++ code, and (b) the s and e on p and r's types force them to be a set and relation, respectively. When a client tries to apply relationalProduct on f and g (line 10) a typing error is raised at compile-time, because the encoding of g is s, rather than e, the expected encoding.

As a third example, we present the function Union_QQ(p,q) from MDDL, which computes the union set of two quasi-reduced BDDs p and q and requires them to have the same number of levels. A portion of the C++ code of this function is shown in Fig. 4 left (lines 1–9); the ASSERTs are used to check that both the input BDDs p and q, and the output BDD, answer, have the same number of levels, k. The client (lines 10–16) tries to compute the union of f and g, two BDDs with different numbers of levels. The library generates a runtime error because the ASSERT on line 3 fails. On the right side of Fig. 4 we show the BDDL equivalent of the library and client code; the l on union's type forces the input BDDs to have the same number of levels and the output to have at at most l levels. The application of union to f and g, which have 1 and 2 levels, respectively, is ill-typed and results in a compile-time error.

As a fourth example, we show how reordering a terminal-only BDD leads to a nullPointerException in JavaBDD. The left side of Fig. 5 shows a simple method, testCrash, that we wrote as an addition to the JavaBDD test suite. The method first performs some initialization/sanity checks (lines 3–4), then creates a BDD, containing just the terminal $\boxed{1}$, and invokes reorder, which generates a nullPointerException. Reordering a terminal-only BDD should not be allowed, or at most should be a no-op. On the right side we show the BDDL code for this scenario; the type of reorder (lines 2–3) stipulates that it can only be invoked on BDDs with at least one level. The application of reorder to $\boxed{1}$ (line 5) is ill-typed and will be rejected, because $\boxed{1}$ is at level 0.

In the companion technical report [14] we provide more examples of how BDDL can be used to specify interfaces of CUDD, MDDL and JavaBDD libraries.

4   The BDDL Language

We now present BDDL, our core language for BDDs. We designed BDDL to support abstractions for key BDD operations, functionality, and semantic properties, based on our examination of several mature decision diagram libraries. BDDL provides forms to build a BDD (via Bnode (...)), use a BDD t (t.level, t.index, t.var, t.tchild, t.fchild), and types, possibly with refinements, to capture semantic properties of BDDs ($bdd[l, r, c]$).

The syntax is shown in Fig. 6. *Indexes* i are unique id's associated with each BDD node: ID_0 and ID_1 are reserved for terminals while id_2, id_3, and higher are used for non-terminals; index uniqueness is a enforced by our typing system. *Strings* g are used to hold variable names; for simplicity, we only allow $x_1, x_2, x_3, \ldots$ as indexes. The *level* l specifies the level of BDD nodes; it can be $\perp$

$$
\begin{array}{lll}
i := & id_2, id_3, id_4, \ldots & index \\
g := & x_1, x_2, x_3, \ldots & string \\
l := & \bot,\; nv & level \\
r := & \bot,\; \mathbf{f},\; \mathbf{q} & \text{fully or quasi reduced} \\
c := & \bot,\; \mathbf{s},\; \mathbf{e} & \text{set or relation} \\
\pi := & \nu \mid l \mid r \mid c & \text{predicate variable} \\
\Gamma := & ID_0 : Id, ID_1 : Id \mid \Gamma, x : \tau \mid \Gamma, h : \tau & \text{typing context} \\
\mu := & \emptyset \quad\mid\quad \mu, h \mapsto v & \text{location binding}
\end{array}
$$

$t ::=$ **Terms**

$$
\begin{array}{lll}
& v & value \\
\mid & x & variable \\
\mid & \mathtt{succ}\ t \quad\mid\quad \mathtt{pred}\ t & \text{successor, predecessor} \\
\mid & \mathtt{iszero}\ t & \text{zero test} \\
\mid & t\ t & application \\
\mid & \mathtt{letrec}\ x : \tau\ \mathtt{=}\ t\ \mathtt{in}\ t & \text{(recursive) let} \\
\mid & \mathtt{if}\ t\ \mathtt{then}\ t\ \mathtt{else}\ t & \text{if statement} \\
\mid & \mathtt{ref}\ t \quad\mid\quad \mathtt{!}t & \text{reference, dereference} \\
\mid & \mathtt{Bnode}\ (t,\ i,\ t,\ t,\ t) & \text{BDD node} \\
\mid & \mathtt{t.level} & \text{level of a node} \\
\mid & \mathtt{t.index} & \text{index of a node} \\
\mid & \mathtt{t.var} & \text{variable of a node} \\
\mid & \mathtt{t.tchild} \mid \mathtt{t.fchild} & \text{true, false child of a node}
\end{array}
$$

$nv ::=$ **Numeric values**

$$
\begin{array}{lll}
& 0 & \text{zero constant} \\
\mid & S(nv) & \text{successor value}
\end{array}
$$

$v ::=$ **Values**

$$
\begin{array}{lll}
& \mathtt{true} \quad\mid\quad \mathtt{false} & \text{boolean value} \\
\mid & nv & \text{numeric value} \\
\mid & i \quad\mid\quad g & \text{index, string} \\
\mid & \lambda x : \tau.t & \text{abstraction value} \\
\mid & h & \text{heap location} \\
\mid & \boxed{0} \quad\mid\quad \boxed{1} & \text{boolean terminal} \\
\mid & \mathtt{Bnode}(v, v, v, v, v) & \text{boolean node}
\end{array}
$$

$\tau ::=$ **Types**

$$
\begin{array}{lll}
& \mathit{bool} \quad\mid\quad \mathit{nat} & \text{booleans, naturals} \\
\mid & \mathit{string} \quad\mid\quad \mathit{Id} & \text{strings, identifiers} \\
\mid & \tau \rightarrow \tau & \text{function type} \\
\mid & \mathit{ref}\ \tau & \text{reference type} \\
\mid & \mathit{bdd}[l, r, c] & \text{node types} \\
\mid & \{\nu : \tau \mid p(\pi)\} & \text{refined type}
\end{array}
$$

$exp ::=$ **Arithmetic expressions**

$$
\begin{array}{lll}
& nv \mid \quad exp + exp \quad\mid\quad exp - exp
\end{array}
$$

$p(\pi) ::=$ **Refinement predicates**

$$
\begin{array}{lll}
& \pi \quad = \quad exp & \text{constant type} \\
\mid & \pi \leq exp \mid \pi \geq exp \mid \pi \neq exp & \text{restrained type} \\
\mid & p(\pi) \quad \wedge \quad p(\pi) & \text{conjunction}
\end{array}
$$

Fig. 6. BDDL syntax.

(unspecified level), or a natural number nv. The *reduction* r of a BDD can be $\bot$ (unspecified), $\mathbf{f}$ (fully reduced), or $\mathbf{q}$ (quasi reduced). The *encoding* c of a BDD can be $\bot$ (unspecified), $\mathbf{s}$ (set), or $\mathbf{e}$ (relation). *Predicate variables* π are used to construct the predicates on refinement types—either terms ν, or variables l, r, or c, which are universally quantified over their respective domains and $\bot$.

The typing context (environment) Γ contains variable names and their associated types $(x : \tau)$, as well as heap locations and their types $(h : \tau)$. We augment Γ with the unique indexes ID_0 and ID_1 assigned to $\boxed{0}$ and $\boxed{1}$, respectively.

The heap μ is a map from references h to values v.

A BDDL *term* can be a value v; a variable x; the successor or predecessor of another term, whose semantics is the successor and predecessor of natural numbers, e.g., the term succ S(S(S(0))), successor of 3 in other words, reduces to $S(S(S(S(0))))$ and corresponds to numerical value 4, whereas pred S(0) reduces to 0, meaning the predecessor of 1 is 0; an iszero test for zero; an application; a letrec binding that allows recursion; an if statement; a reference ref t or dereference !t. Note the presence of references, but the lack of assignment in our system: this is by choice, as we want to allow sharing BDD nodes but avoid mutation to preserve a BDD's structural integrity. We represent BDDs via 5-tuples, i.e., terms Bnode(t_1,i,t_3, t_4,t_5); a Bnode is a boolean node in a BDD; t_1 is the node level (a natural number), i represents the index of the node (a unique identifier), t_3 holds the variable associated with a node (a string), t_4 represents a reference to the true ('1') child of the node and t_5 is a reference to its false ('0') child. To extract the elements of the 5-tuple, we use t.index, t.level, t.var, t.tchild, and t.fchild; i.e., (Bnode(t_1,i,t_3,t_4,t_5)).level= t_1, (Bnode(t_1,i,t_3,t_4,t_5)).index = i, and so on.

The companion technical report [14] illustrates the use of the BDDL language to model a BDD with 3 levels. The whole BDD is represented by t_6=Bnode(3,id_6, x_2,t_4,t_5) since t_4 and t_5 are references to the children that are themselves Bnode terms, i.e., t_4 contains information about its children t_2 and t_1 which in turn contain information about their children. In this case, we reach the terminal nodes, hence we have the information about the entire diagram. We also show the BDDL types of these terms, described in the next section. Recall that we use t.level, t.index, t.var, t.tchild and t.fchild to extract components of a Bnode. For the previous example, we have t.level(id_6) = 3, t.index(id_6) = id_6, t.tchild(id_6) = id_4, and t.fchild(id_6) = id_5.

Values denote the possible final results of an evaluation. A value in BDDL is either a boolean constant true or false; a string or an identifier; the number 0 or a non-zero natural number $S(nv)$; an abstraction $\lambda x : \tau.t$; a heap location h; BDD terminal nodes $\boxed{0}$ or $\boxed{1}$, or a non-terminal node Bnode(v_1, v_2, v_3, v_4, v_5) where v_1, v_2, v_3, v_4, and v_5 are values. The typing system, and the syntax of refinements are defined next.

5   Typing, Semantics, and Soundness

5.1   Types

We use primitive types *bool* and *nat* to denote the sets of boolean values and natural numbers, respectively; *string*s are used only for representing variable

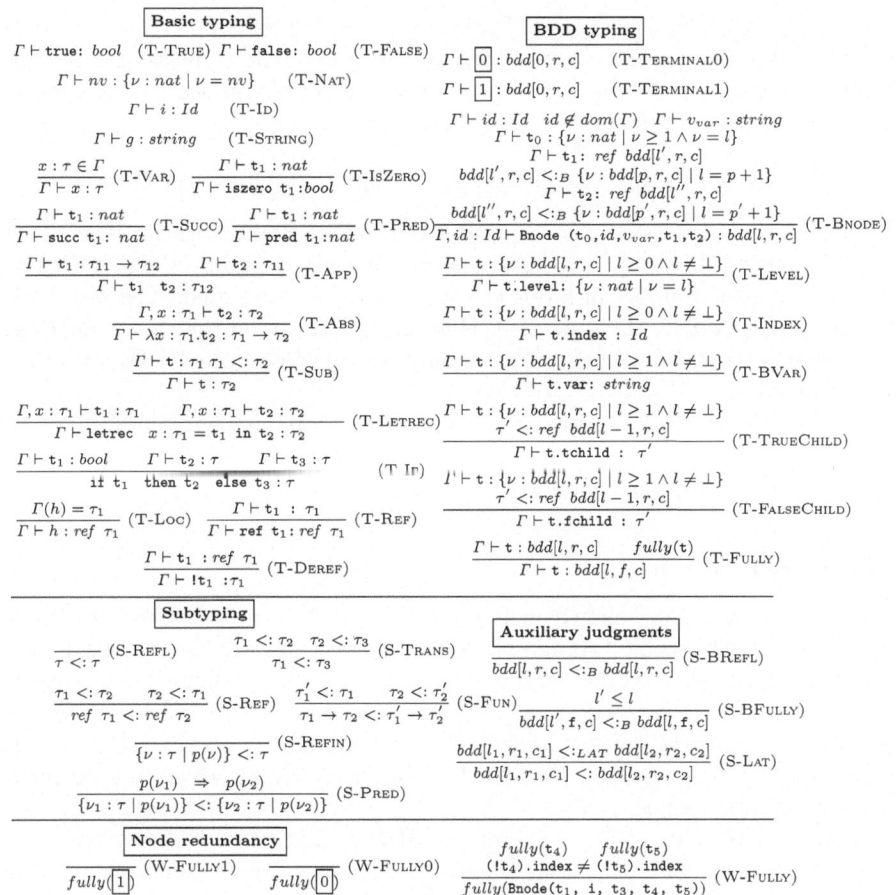

Fig. 7. BDDL typing.

names (accessible via .var) in BDD nodes. We use Id to represent identifier types; we define this as a type distinct from $string$ or nat to account for different DD implementations using different representations (e.g., int, string); the only values that inhabit this type are the unique id's associated with BDD nodes (accessible via .index). Function types have the standard representation, $\tau \rightarrow \tau$, as do references, $ref\ \tau$.

$bdd[l, r, c]$ is the fundamental type in our language. A BDD node has type $bdd[l, r, c]$, where l refers to the level of the node, r refers to the reduction form of the decision diagram, and c refers to the encoding of the diagram (a set or a relation). For example, node t_6 = Bnode(3, id_6, x_2, t_4, t_5) ([14]) has type $bdd[3, r, c]$, i.e., it is a node at level 3 of a BDD with no restriction on the reduction or the encoding.

To express semantic BDD properties, we use refinement types [8], in the form proposed by Rondon et al. [19]. A refined type $\{\nu : \tau \mid p(\pi)\}$ expresses a refinement of the primitive type τ. For example, the refined type $\{\nu : nat \mid \nu \geq 1\}$ describes the set of natural numbers that are greater or equal to 1; the type $\{\nu : nat \mid \nu = n\}$ describes the type of the natural number n; the type of function Union_QQ from Fig. 4 in Sect. 3 is:

$$bdd[l, \mathsf{q}, \mathsf{s}] \rightarrow bdd[l, \mathsf{q}, \mathsf{s}] \rightarrow \{\nu : bdd[l', \mathsf{q}, \mathsf{s}] \mid l' \leq l\})$$

meaning it takes two quasi-reduced set-encoding BDDs with arbitrary (but equal) numbers of levels l, and returns a quasi-reduced set-encoding BDD with at most the same number of levels l. Note how r and c are quantified over their domains; for instance, the type of $\boxed{1}$ is $bdd[0, r, c]$, meaning that a terminal can only be at level 0, but can be part of fully- or quasi-reduced BDDs, encoding sets or relations.

Finally, the language of predicates $p(\pi)$ used on refinement types allows us to express equality and inequality refinements, e.g., $\{\nu : nat \mid \nu = 2\}$, and conjunctions of simple arithmetic expressions, e.g., $\{\nu : nat \mid \nu \geq 1 \wedge \nu = 5\}$.

5.2   Typing Rules

The BDDL typing rules shown in Fig. 7 can be split into two categories: basic typing and BDD-specific typing. The basic typing rules are the standard rules for lambda calculus extended with booleans and unary representation of natural numbers, as presented by Pierce [18], with the following modifications: when type-checking natural values nv via (T-Nat), we use refinements to represent their value, i.e., $\{\nu : nat \mid \nu = nv\}$; we add the rules (T-Id) and (T-String) to type-check id's and variable names.

The top right side of Fig. 7 shows our BDD typing rules. Rules (T-Terminal0) and (T-Terminal1) stipulate that the $\boxed{0}$ and $\boxed{1}$ terminals can be found in any BDD at level 0.

(T-Bnode) is the fundamental rule in our system. To enforce identifier uniqueness, we require $id \notin dom(\Gamma)$. We require that $\mathsf{t}_0 : \{\nu : nat \mid \nu \geq 1 \wedge \nu = l\}$ i.e., that the node's level be at least 1 (to ensure that Bnodes are non-terminals), and the same level l as in the conclusion of the judgment. To allow sharing, we store the children t_1 and t_2 by reference; they have types $bdd[l', r, c]$ and $bdd[l'', r, c]$, respectively. We now explain how BNodes can store both fully and quasi-reduced BDDs, and do so safely. Consider the $<:_B$ subtyping premise for the true child:

$$bdd[l', r, c] <:_B \{\nu : bdd[p, r, c] \mid l = p + 1\}$$

If this is a quasi-reduced BDD which allows no level skipping, then we have $l' = p = l - 1$. If this is a fully-reduced BDD which allows level skipping, then we have $l' \leq p$ where $p = l - 1$, and by applying the (S-BFully) subtyping rule, we can actually have $l' < p$; for instance, a Bnode at level 3 can have its children be at level 1; the premises for the false-child t_2 are similar. Note the use of the same variables r and c in both the children and current node (premises and conclusion of the rule) which forces the BDD to be consistent: either fully-reduced or quasi-reduced, and, respectively, either set-encoding or relation-encoding. Note that the $<:_B$ relationship is different from the subtyping relationship, as it only allows us to establish a relationship between reduction forms.

The rule (T-LEVEL) type-checks extracting a node's level. The level is a refined *nat* type (refinement: $l \geq 0 \wedge l \neq \perp$) with the same l as in the refinement of the BDD node; we allow $l = 0$ because asking for the level of a terminal is permitted; we do not allow $l = \perp$ because $\perp$ means unspecified level. The rule (T-INDEX) is similar—we can ask for the level of any node (including terminals, whose indexes are ID_0 and ID_1) as long as its level is not $\perp$. The rule (T-BVAR) is similar, though it only works on non-terminals ($l \geq 1$), as terminals cannot encode variables. The rule (T-TCHILD) type-checks the extraction of the true-child; the premises are the same as for (T-BVAR), since we can only extract children of non-terminals; note again the subtyping in the premises, which allows level skipping for fully-reduced, but not for quasi-reduced, BDDs; (T-FCHILD) is similar.

Our system lacks type polymorphism, which we omit for simplicity at the expense of expressivity. Note however that the variables l, r and c used on type rules allow specification of quasi-reduced, fully-reduced, or generic BDDs. For example, we can construct BDDs whose reducing and encoding are unspecified, and allow functions to operate on them (albeit the range of operations is constrained, as with any polymorphic function). Consider a simple BDD, BNode($1,id_2,x_1$,ref $\boxed{0}$,ref $\boxed{1}$), that contains one non-terminal node with two children, $\boxed{0}$ and $\boxed{1}$; its type is $\{\nu : bdd[l,r,c] \mid l = 1\}$, or, in short, $bdd[1,r,c]$. This BDD is "generic", in that it can be safely attached to both a fully- or quasi-reduced BDD, that encodes either a set or a relation. Then the BDD can be safely used in a concrete context, e.g., with reduction f and an encoding s.

To express the absence of redundant nodes in fully-reduced BDDs, we introduce the rules (W-FULLY0), (W-FULLY1), (W-FULLY), and (T-FULLY). The first two rules express the non-redundancy of terminal nodes—they are unique and do not have children nodes. A fully-reduced BDD must not contain any redundant nodes, i.e., its outgoing edges must not point to the same node. We capture this property by applying (W-FULLY) recursively and by checking at each recursion that the indexes of the children nodes are different (this prevents a node from being redundant by prohibiting the true edge and the false edge from pointing to the same node) and that they are fully-reduced. Finally, (T-FULLY) prevents passing a BDD with the type $bdd[l,r,c]$ when a BDD with type $bdd[l,\mathtt{f},c]$ is expected when the BDD is not fully-reduced.

5.3   Subtyping

Subtyping is essential in our system, as it allows types that contain more information about a BDD node (e.g., fixed number of levels) to be used in contexts where knowing the number of levels is not required; also, it allows refined types with strong conditions to be used in contexts where types' conditions are more relaxed. The subtyping rules are shown on the bottom of Fig. 7. (S-REFL), (S-REF), (S-TRANS), and (S-FUN) are standard. (S-BFULLY) allows level skipping for fully-reduced BDDs and (S-BREFL) allows the B-subtyping relationship to be reflexive. Rule (S-REFIN) indicates that a refinement of a type τ is a subtype of τ. Rule (S-PRED)

allows us to use a more constrained refinement type, i.e., with a stronger predicate $p(\nu_1)$ (e.g., $\nu \geq 0 \wedge \nu \neq 3$) in a context that only requires a weaker predicate
$p(\nu_2)$ (e.g., $\nu_2 \geq 1$). Note that this rule is only used for ν predicates but not for
the other predicates since weakening the l, r and c constraints in a $bdd[l, r, c]$ has
a different semantics. As an example, $\{\nu : bdd[l, r, c] \mid l = 0\}$ is not a subtype of
$\{\nu : bdd[l, r, c] \mid l \geq 0\}$ even though $(l = 0) \Rightarrow (l \geq 0)$ since a terminal node can
only be at level 0.

(S-LAT) stipulates that a BDD with type $bdd[l_1, r_1, c_1]$ can be used in a context
where a $bdd[l_2, r_2, c_2]$ is required, as long as $bdd[l_1, r_1, c_1]$ is on a downward path
from $bdd[l_2, r_2, c_2]$ in the subtyping lattice. Figure 8 shows the lattice of subtyping
relationships in our system; we use $\perp$ to represent unspecified values. Essentially,
$\perp$ is a union type over the domains of l, r, and c.

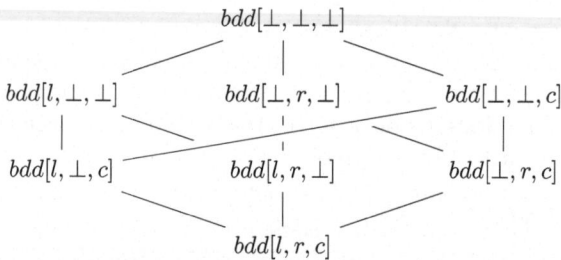

Fig. 8. Subtyping lattice (the $<:_{LAT}$ relation).

In the companion technical report [14] we provide two examples of how type
checking and inference are performed in BDDL.

5.4   Semantics and Soundness

The operational semantics is standard, small-step, defined as reduction on terms
t and memory stores μ, i.e.,

$$\mu; t \longrightarrow \mu'; t'$$

The full semantics can be found in the companion technical report [14]. We
now state the progress and preservation lemmas for our system.

Lemma 1 (Progress). *If t is a closed, well-typed term (such that $t : \tau$) then
either t is a value or else, for any store μ such that $\Gamma \vdash \mu$, there exists a term
t' and store μ' with $\mu; t \longrightarrow \mu'; t'$*

Proof. By induction on the typing derivation $t : \tau$.

Lemma 2 (Preservation). *If $\Gamma \vdash t : \tau, \Gamma \vdash \mu$ and $\mu; t \longrightarrow \mu'; t'$ then for some
Γ', we have $\Gamma' \vdash t' : \tau$ and $\Gamma' \vdash \mu'$.*

Proof. By induction on the typing derivation $\Gamma \vdash t : \tau$.

We can now state the soundness theorem: well-typed programs do not go wrong.

Theorem 1 (Soundness). *If $\Gamma \vdash t : \tau$, and $\Gamma \vdash \mu$, then either t is a value, or there exist Γ', μ', t', such that $\mu ; t \longrightarrow \mu' ; t'$ and $\Gamma' \vdash t' : \tau$ and $\Gamma' \vdash \mu'$.*

6   Related Work

The work closest to ours, on which we partially build, is Kawaguchi et al. [13]'s liquid types (which in turn built upon Rondon et al.'s work [19]). We borrow their syntax of refinement types expressed as logically quantified predicates over program variables. In addition, we add formal support for expressing and checking structural constraints on BDDs. They can check a broad-range of properties on various data structures, such as lists, trees, heaps, maps, and vectors. In contrast, we focus on one data structure, BDD, but express and verify a broader range of structural and logical properties. For example, they can check BDDs for ordering: mk_not::x:bdd→ $\{\nu : bdd|\ var\ x \leq var\ \nu\}$.
The same type can be expressed in our formalism as:
mk_not:$\{\nu_1 : bdd[l,r,c]\} \rightarrow \{\nu_2 : bdd[l',r',c'] \mid \nu_1.\texttt{var} \leq \nu_2.\texttt{var}\}$.
However, this type does not guarantee preservation of structural integrity, such as number of levels, reduction, or encoding status. We believe that their system can capture some of these properties but, as our focus is on BDDs, we can express the properties more directly and concisely. For instance, the type of reorder (with the natural order $x_1 \leq x_2 \leq \cdots \leq x_n$) is, in our system:
$\{\nu_1 : bdd[l, r, c] \mid l \geq 1\} \rightarrow \{\nu_2 : bdd[l, r, c] \mid \nu_2.\texttt{var} \leq\ !(\nu_2.\texttt{tchild}).\texttt{var} \wedge \nu_2.\texttt{var} \leq\ !(\nu_2.\{\texttt{fchild}\}).\texttt{var}\}$, which enforces that the levels, reduction, and encoding of input BDD and output BDD are the same, hence preserving structural integrity. Our system is less expressive in general, as we do not allow type-level polymorphism or nested refinements. We have pursued one data structure, BDD, and its manipulation in three DD libraries; they have verified the implementations of a wide-range of data structures from OCaml's standard library.

Drechsler [6] presented a run-time technique for BDD verification, using a recursive checksum method to verify BDD integrity. Their approach has been motivated by memory errors, e.g., copy faults and errors due to aliasing; in their approach, such memory errors are detected at runtime. Our approach is intended to capture and verify higher-level properties, and to do so at compile time; we do not have a memory model, though we can prevent errors such as use-before-allocation, because we do not allow uninitialized references.

Stanković and Astola [21] take a syntactic approach to checking the structural integrity of BDDs. A Treetype XML wrapper stores information about a particular BDD instance (e.g., number of variables, levels, edges) as attribute values, and these values are used to reason about what operations are permitted for this particular BDD instance—the BDD's body and collection of attributes constitute a type which can be type-checked by an XML parser. Their approach can be readily extended to more elaborate decision diagram flavors, such as EVBDD, by merely adding attribute fields. In contrast to this syntactic approach, our method uses a type system and semantics to check BDD-manipulating code, so BDDs created by BDDL-checked code are correct by construction.

Bernasconi et al. [4] as well as Bernasconi and Ciriani [3] have proposed dynamic approaches for detecting and repairing index or pointer errors in Ordered Binary Decision Diagrams. Their approach differs from ours in the same way as described for the previous two approaches, i.e., our approach is enforcing a different set of properties, and does so statically.

Giorgino and Strecker [10] as well as Ortner and Schirmer [16,17] provide mechanized proofs in Isabelle for Isabelle/HOL-expressed implementations, e.g., proving certain BDD properties by casting the verification of BDD (normalization) as a particular case of verifying pointer-manipulating programs. Our goal was not a mechanized proof, but rather BDDL is a type system for static checking. For example a CUDD checker, or a checker for any substantial implementation written in popular imperative languages, could be built by adding a simple front-end that maps C++ to BDDL. It is unclear how a substantial imperative implementation such as CUDD can be transformed into Isabelle to be checked.

7   Conclusion

We have presented BDDL, a calculus to verify BDD correctness. BDDL combines language support to build and safely manipulate BDDs with refinement types that allow programmers to concisely express structural and logical invariants. We formalized BDDL using a type system and small-step operational semantics, and proved it sound. We have presented examples of how frequently-used BDD library functions can be expressed and statically verified using BDDL. We plan to extend this work in two directions: (1) automatic verification of library code, and (2) extend BDDL to reason about, and verify, properties of more general variants of decision diagrams, such as multi-way, multi-terminal, and edge-valued.

References

1. Miner, A., et al.: MEDDLY: multi-terminal and Edge-valued Decision Diagram LibrarY. https://meddly.sourceforge.io/
2. Bergman, D., Cire, A.A., van Hoeve, W.J., Hooker, J.: Decision Diagrams for Optimization. Springer, Cham (2016). https://doi.org/10.1007/978-3-319-42849-9
3. Bernasconi, A., Ciriani, V.: Index-resilient zero-suppressed bdds: definition and operations. ACM Trans. Des. Autom. Electron. Syst. **21**(4), 1–27 (2016)
4. Bernasconi, A., Ciriani, V., Lago, L.: On the error resilience of ordered binary decision diagrams. Theor. Comput. Sci. **595**, 11–33 (2015)
5. Bryant, R.: On the complexity of VLSI implementations and graph representations of Boolean functions with application to integer multiplication. IEEE Trans. Comput. **40**(2), 205–213 (1991)
6. Drechsler, R.: Verifying integrity of decision diagrams. Integr. **32**(1–2), 61–75 (2002)

7. Somenzi, F.: CUDD: CU Decision Diagram Package. https://github.com/ivmai/cudd

8. Freeman, T., Pfenning, F.: Refinement types for ML, pp. 268–277. PLDI 1991, Association for Computing Machinery, New York, NY, USA (1991)

9. Ciardo, G., Miner, A.S.: SMART: Stochastic model-checking analyzer for reliability and timing. https://asminer.github.io/smart/

10. Giorgino, M., Strecker, M.: Correctness of pointer manipulating algorithms illustrated by a verified BDD construction. In: Giannakopoulou, D., Méry, D. (eds.) FM 2012. LNCS, vol. 7436, pp. 202–216. Springer, Heidelberg (2012). https://doi.org/10.1007/978-3-642-32759-9_18

11. Groen, F., Smidts, C., Mosleh, A., Swaminathan, S.: Qras - the quantitative risk assessment system. In: Annual Reliability and Maintainability Symposium. 2002 Proceedings (Cat. No.02CH37318), pp. 349–355 (2002)

12. Whaley, J.: JavaBDD. http://javabdd.sourceforge.net/

13. Kawaguchi, M., Rondon, P., Jhala, R.: Type-based data structure verification, pp. 304–315. PLDI 2009 (2009)

14. Lembachar, Y., Rusich, R., Neamtiu, I., Ciardo, G.: BDDL: a type system for binary decision diagrams. Technical Report, Department of Computer Science, NJIT, May 2022. https://web.njit.edu/~ineamtiu/pubs/bddl-tr.pdf

15. Loekito, E., Bailey, J., Pei, J.: A binary decision diagram based approach for mining frequent subsequences. Knowl. Inf. Syst. 24(2), 235–268 (2010)

16. Ortner, V., Schirmer, N.: Verification of BDD normalization. In: Hurd, J., Melham, T. (eds.) Theorem Proving in Higher Order Logics, pp. 261–277 (2005)

17. Ortner, V., Schirmer, N.: Bdd normalisation. Archive of Formal Proofs, Febuary 2008. https://isa-afp.org/entries/BDD.html, Formal proof development

18. Pierce, B.C.: Types and Programming Languages. MIT Press, Cambridge (2002)

19. Rondon, P.M., Kawaguchi, M., Jhala, R.: Liquid types, pp. 159–169. PLDI 2008, June 2008

20. Siminiceanu, R.I., Ciardo, G.: Formal verification of the Nasa runway safety monitor. Int. J. Softw. Tools Technol. Transf. 9(1), 63–76 (2007)

21. Stanković, S., Astola, J.: Xml framework for various types of decision diagrams for discrete functions. IEICE Trans. 90-D, 1731–1740 (2007)

22. Whaley, J., Lam, M.S.: Cloning-based context-sensitive pointer alias analysis using binary decision diagrams, PLDI 2004, pp. 131–144 (2004)

23. Xing, L., Amari, S.V.: Binary Decision Diagrams and Extensions for System Reliability Analysis. Wiley, Hoboken (2015)

24. Yanushkevich, S.N., Miller, D.M., Shmerko, V.P., Stankovic, R.S.: Decision Diagram Techniques for Micro- and Nanoelectronic Design Handbook (2006)

25. Yoon, S., De Micheli, G.: An application of zero-suppressed binary decision diagrams to clustering analysis of DNA microarray data, EMBC 2004, pp. 2925–2928 (2004)

Definitional Quantifiers Realise Semantic Reasoning for Proof by Induction

Yutaka Nagashima[✉] [iD]

Independent Researcher, Cambridge, UK
united.reasoning@gmail.com

Abstract. Proof assistants offer tactics to apply proof by induction, but these tactics rely on inputs given by human engineers. To automate this laborious process, we developed SeLFiE, a boolean query language to represent experienced users' knowledge on how to apply the induct tactic in Isabelle/HOL: when we apply an induction heuristic written in SeLFiE to an inductive problem and arguments to the induct tactic, the SeLFiE interpreter judges whether the arguments are plausible for that problem according to the heuristic by examining both the syntactic structure of the problem and definitions of the relevant constants. To examine the intricate interaction between syntactic analysis and analysis of constant definitions, we introduce *definitional quantifiers*. For evaluation we build an automatic induction prover using SeLFiE. Our evaluation based on 347 inductive problems shows that our new prover achieves $1.4 \cdot 10^3\%$ improvement over the corresponding baseline prover for 1.0 s of timeout and the median value of speedup is 4.48x.

1 Introduction

The automation of proof by induction is a long standing challenge in Computer Science. Conventionally, human researchers manually investigate both inductive problems and relevant definitions to decide how to apply proof by induction. To mechanise such analysis, this paper introduces *definitional quantifiers*: quantifiers that range over the defining clauses of relevant constants to capture semantic properties of inductive problems.

1.1 Motivating Example

Consider the following two ways to define a reverse function for lists presented in a tutorial of Isabelle/HOL [39]:

```
@ :: α list ⇒ α list ⇒ α list
   [] @ ys = ys
| (x # xs) @ ys = x # (xs @ ys)

rev1 :: α list ⇒ α list
```

© The Author(s), under exclusive license to Springer Nature Switzerland AG 2022
L. Kovács and K. Meinke (Eds.): TAP 2022, LNCS 13361, pp. 48–66, 2022.
https://doi.org/10.1007/978-3-031-09827-7_4

```
  rev1 [] = []
| rev1 (x # xs) = rev1 xs @ [x]

rev2 :: α list ⇒ α list ⇒ α list
  rev2 [] ys = ys
| rev2 (x # xs) ys = rev2 xs (x # ys)
```

where # is the list constructor, [x] is a syntactic sugar for x # [], and @ is
the infix operator to append two lists into one. How do you prove the following
equivalence lemma?

```
lemma "rev2 xs ys = rev1 xs @ ys"
```

Since both reverse functions are defined recursively, it is natural to guess we
can tackle this problem with proof by induction. But how do you apply proof by
induction to this inductive problem? In this paper, we present SeLFiE, a boolean
query language to encode induction heuristics in a declarative form, and its fast
interpreter developed from scratch. SeLFiE is embedded in Isabelle/ML, the
implementation language of Isabelle/HOL, and implemented for Isabelle2020.
The key idea behind SeLFiE is *definitional quantifiers*: new kinds of quantifiers
that allow for definitional reasoning in a domain-agnostic style.

1.2   Background

A prominent proof automation approach for proof assistants is the so-called
hammer-style tools, such as HOL(y) Hammer [19] for HOL-light [13], CoqHam-
mer [9] for Coq, and Sledgehammer [2] for Isabelle/HOL [39]. Sledgehammer,
for example, translates proof goals in the polymorphic higher-order logic of
Isabelle/HOL to monomorphic first-order logic and attempts to prove the translated goals using various external automated provers. Even though Sledgehammer brought powerful automation to Isabelle/HOL [3]; when it comes to inductive theorem proving the essence of inductive problems is lost in the translation,
severely impairing the performance of Sledgehammer.

This is unfortunate: most analyses of programs and programming languages
involve reasoning about recursive data structures and procedures containing
recursion or iteration [6], and inductive problems are essential to these analyses.

We address this long standing challenge with SeLFiE. SeLFiE stands for
semantic-aware logical feature extraction. SeLFiE has two main features: *definitional quantifiers*, and *domain-agnosticism*. Domain-agnosticism allows users to
encode induction heuristics that can transcend problem domains, whereas definitional quantifiers allow SeLFiE heuristics to examine not only the syntactic
structures of inductive problems but also the definitions of relevant constants.

Our implementation, available at GitHub [30], is specific to Isabelle/HOL:
we implemented our system as an Isabelle theory for smooth user experience.
However, the underlying concept of definitional reasoning is transferable to other

proof assistants, such as Coq, Lean [29], and HOL [46]: no matter which proof assistant we use, we need to reason over not only the syntactic structure of proof goals but also definitions relevant to the goals to decide how to apply proof by induction.

The rest of the paper is organized as follows. Section 2 shows how to apply proof by induction in Isabelle using the example from Sect. 1.1 and clarifies the need for reliable heuristics. Section 3 gives an overview of what we mean by encoding induction heuristics and applying them to inductive problems in Isabelle/HOL. Since it is still a new approach to reason over inductive problems using a boolean query language, Sect. 4 reviews LiFtEr [31], an existing framework developed to encode syntax-based heuristics for Isabelle/HOL. In particular, we observe how LiFtEr's quantifiers allow us to write heuristics in a domain-agnostic style. Then, we identify what induction heuristics we can*not* encode in LiFtEr. In Sect. 5, we present SeLFiE and its fast interpreter developed from scratch. In addition to the domain-agnosticism given by LiFtEr's quantifiers, SeLFiE enables definitional reasoning using new language constructs that allow for the reasoning about both the syntactic structure of proof goals and the definitions of relevant constants. In Sect. 6 we introduce a recommendation system for the induct tactic as a use case of SeLFiE and build a fast automatic inductive prover using this recommendation system, and we discuss how much performance gain SeLFiE brought to inductive theorem proving in Isabelle/HOL.

2   Proof by Induction in Isabelle/HOL

Modern proof assistants come with *tactics* to facilitate proof by induction. For example, Isabelle/HOL offers the induct tactic. The user-interface of the induct tactic allows for an intuitive application of proof by induction. For example, Nipkow *et al.* [39] proved our motivating example as follows:

```
lemma model_proof: "rev2 xs ys = rev1 xs @ ys"
 apply(induct xs arbitrary: ys) by auto
```

That is to say, they firstly applied structural induction on xs while generalizing ys. Since xs is a list of any type, this application of structural induction resulted in the following two sub-goals:

```
1. ∀ys. rev2 [] ys = rev1 [] @ ys
2. ∀a xs ys. (∀ys. rev2 xs ys = rev1 xs @ ys) ⟹
   rev2 (a # xs) ys = rev1 (a # xs) @ ys
```

where $\forall$ and $\Longrightarrow$ represent the universal quantifier and implication of Isabelle's underlying logic respectively. The first sub-goal is the base case for the structural induction, whereas the second sub-goal is the step case where we are asked to prove that this conjecture holds for (a # xs) and ys, assuming that the conjecture holds for the same xs and an *arbitrary* ys. Then, they proved the remaining sub-goals using the general purpose tactic, auto. For the step case, auto rewrote the left-hand side of the meta-conclusion as follows:

```
    rev2 (a # xs) ys                using the second clause defining rev2
↪ rev2 xs (a # ys)
```

whereas auto rewrote the right-hand side as follows:

```
    rev1 (a # xs) @ ys              using the second clause defining rev1
↪ (rev1 xs @ [a]) @ ys                using the associative property of @
↪ rev1 xs @ ([a] @ ys)                using the second clause defining @
↪ rev1 xs @ (a # ([] @ ys))            using the first clause defining @
↪ rev1 xs @ (a # ys)
```

Applying such rewriting, auto internally transformed the step case to the following intermediate goal:

```
∀a xs ys. (∀ys. rev2 xs ys = rev1 xs @ ys) ⟹
  rev2 xs (a # ys) = rev1 xs @ (a # ys)
```

Since ys was generalized in the induction hypothesis, auto proved rev2 xs (a # ys) = rev1 xs @ (a # ys) by considering it as a concrete case of the induction hypothesis. If Nipkow *et al.* had not passed ys to the arbitrary field, the induct tactic would have produced the following sub-goals:

```
1. rev2 [] ys = rev1 [] @ ys
2. ∀a xs ys. (rev2 xs ys = rev1 xs @ ys) ⟹
  rev2 (a # xs) ys = rev1 (a # xs) @ ys
```

This step case requests us to prove that the original goal holds for (a # xs) and ys, assuming that it holds for the same xs and the *same* ys that appear in the induction hypothesis. If we apply auto to these sub-goals, auto proves the base case, but it leaves the step case as follows:

```
∀a xs. rev2 xs ys = rev1 xs @ ys ⟹
  rev2 xs (a # ys) = rev1 xs @ (a # ys)
```

That is, auto is unable to complete the proof attempt because ys is shared both in the conclusion and induction hypothesis, illustrating the importance of variable generalization.

Note that we did not have to develop induction principles manually for model_proof since the induct tactic found out how to apply structural induction from the arguments passed by Nipkow *et al.* In fact, for most of the time Isabelle users do not have to develop induction principles manually, but they only have to pass the right arguments to the induct tactic.

Furthermore, there are often multiple equally appropriate ways to prove one theorem. For example, we could have proved our running example with the following script: apply (induct xs ys rule: rev2.induct) by auto. This script applies computation induction using the auxiliary lemma, rev2.induct, in the rule field. Fortunately, in many cases Isabelle automatically creates such auxiliary lemmas when defining relevant constants. In our case, Isabelle derived rev2.induct automatically when defining rev2. This way, the induct tactic reduces the problem of how to apply induction to the following three questions:

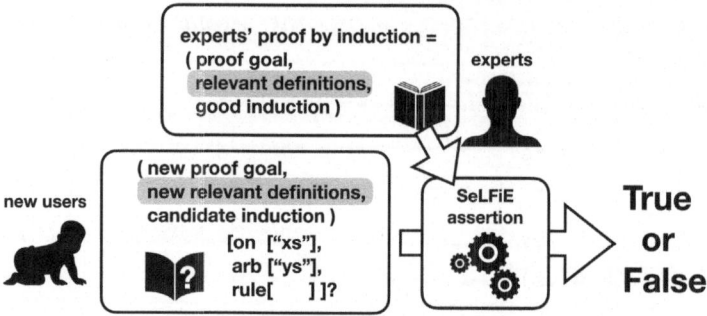

Fig. 1. The overview of SeLFiE.

- On which terms do we apply induction?
- Which variables do we pass to the `arbitrary` field to generalize them?
- Which rule do we pass to the `rule` field?

However, answering these questions is a well-known challenge, which used to require hard-won expertise. We developed SeLFiE to encode such expertise.

3   Overview of SeLFiE

Figure 1 shows how SeLFiE transfers such experienced users' knowledge to new users: when experienced users tackle inductive problems of their own, they encode their expertise about how they use the `induct` tactic as SeLFiE heuristics. Each SeLFiE heuristic is an assertion that takes a triple of a proof goal, relevant constant definitions, and arguments passed to the `induct` tactic. A well-written SeLFiE assertion should return True if the arguments to the `induct` tactic are likely to be useful to prove the problem, whereas it should return False if the combination is not likely to be useful to prove the problem. When new users want to know if their use of the `induct` tactic is appropriate or not, they apply the assertion written by an expert to their own problem and learn if their choice of arguments is compatible with the induction heuristic encoded by the expert. Note that we highlighted parts of Fig. 1 to emphasize the main differences from the SeLFiE's predecessor, LiFtEr, developed for a similar purpose.

Originally, we developed SeLFiE's interpreter as an interactive tool to test a choice of proof by induction in terms of experts' heuristics. However, we can also use SeLFiE to build fully automated inductive provers as shown in Sect. 6. In the following, we review LiFtEr and explain why we need a reasoning framework that can take relevant definitions into account to encode reliable heuristics.

Syntax 1 The abstract syntax of LiFtEr / SeLFiE in one. The language components unique to SeLFiE are highlighted.

---

argument := *term* | *number*
literal := *term_occ* | *rule* | *argument* | ...
assertion := *atomic* | *literal* | *connective* | *quantifier* | (*assertion*)
         | λ *assertions*. *assertion* | *assertion assertions*
type := term | term_occ | rule | number
modifier := induction | arbitrary | rule
quantifier := ∃x : *type*. *assertion*
         | ∀x : *type*. *assertion*
         | ∃x : *term* ∈ *modifier*. *assertion*
         | ∀x : *term* ∈ *modifier*. *assertion*
         | ∃x : *term_occ* ∈ y : term. *assertion*
         | ∀x : *term_occ* ∈ y : term. *assertion*
         | ∃$_D$(*term* , λ *arguments*. *assertion* , *arguments*)
         | ∀$_D$(*term* , λ *arguments*. *assertion* , *arguments*)
connective := True | False | *assertion* ∨ *assertion* | *assertion* ∧ *assertion*
 | *assertion* → *assertion* | ¬ *assertion*
atomic := term_is_free (*term*)
       | are_same_term (*term* , *term*)
       | is_nth_argument_of (*term_occ*, *number*, *term_occ*)
       | is_nth_argument_in (*term_occ*, *number*, *term_occ*)
       | are_of_same_term (*term_occ* , *term_occ*) | ...

---

4 Syntactic Reasoning in LiFtEr

4.1 LiFtEr: Logical Feature Extraction

LiFtEr is the first framework designed to describe how to use the induct tactic without relying on domain-specific constructs. Syntax 1 outlines LiFtEr's syntax, which resembles that of first-order logic. When reading Syntax 1, we ignore highlighted parts, which we discuss in Sect. 5.1.

As shown in Syntax 1, LiFtEr offers four primitive variable types: natural numbers, induction rules, terms, and term occurrences. An induction rule is an auxiliary lemma passed to the rule field of the induct tactic. The domain of terms is the set of all sub-terms appearing in the inductive problem at hand, whereas the domain of term occurrences is the set of all occurrences of such sub-terms. LiFtEr distinguishes terms and term occurrences explicitly because we often have multiple distinct occurrences of the same term in a syntax tree and have to analyze the locations of such occurrences. For instance, the variable ys appears twice in our theorem about list reversal. But what matters when deciding which variables to generalize is the occurrence of ys on the left-hand side and its location relative to the only occurrence of rev2, as we shall see in Sect. 5.2. Quantifiers over terms can be restricted to those terms that appear as arguments to the induct tactic under consideration.

Program 1 Naive generalization heuristic in LiFtEr

∀ *free_var* : term.
  term_is_free (*free_var*)
  ∧
    ¬ ∃ *induct* : induction. are_same_terms (*free_var*, *induct*)
  ⟶
    ∃ *generalized* : arbitrary. are_same_terms (*free_var*, *generalized*)

4.2 Naive Generalization Heuristic in LiFtEr

As we saw in Sect. 2, the key to the successful application of the induct tactic for our motivating example is the generalization of ys using the arbitrary field. When explaining why they decided to generalize ys, Nipkow *et al.* introduced the following generalization heuristic [38]:

> Generalize induction by generalizing all free variables (except the induction variable itself).

We can encode this generalization heuristic in LiFtEr as shown in Program 1. In plain English, Program 1 reads as follows:

> For any term, *free_var*, in a proof goal, if *free_var* is a free variable but not passed to the induct tactic as an induction term, there exists a term, *generalized*, in the arbitrary field such that *free_var* and *generalized* are the same term.

If we evaluate this heuristic for our ongoing example and its model proof by Nipkow *et al.*, the LiFtEr interpreter returns True, approving the generalization of ys. But this heuristic seems too coarse to produce reliable recommendations. In fact, Nipkow *et al.* articulate the limitation of this heuristic:

> However, it (this generalization heuristic) should not be applied blindly. It is not always required, and the additional quantifiers can complicate matters in some cases. The variables that need to be quantified are typically those that change in recursive calls.

Unfortunately, it is not possible to encode this provision in LiFtEr because it involves reasoning on the structure of the syntax tree representing the definition of a constant appearing in a proof goal, which is rev2 in this particular case. In other words, LiFtEr heuristics can describe the structures of proof goals in a domain-independent style, but they cannot describe the structures of relevant constants' definitions. What is much needed is a framework to reason about both arbitrary proof goals and their relevant definitions in terms of the arguments passed to the induct tactic in a domain-agnostic style. And this is the main challenge addressed by SeLFiE.

5   Semantic Reasoning in SeLFiE

5.1   Semantics-Aware Logical Feature Extraction

We designed SeLFiE to overcome LiFtEr's limitation while preserving its capability to transcend problem domains. Syntax 1 presents the abstract syntax of SeLFiE. Since SeLFiE inherits design choices from LiFtEr, we re-use Syntax 1; however, we now include the highlighted constructs into our consideration.

Compared to LiFtEr, which resembles first-order logic, SeLFiE adopts lambda abstractions and function applications to support the *definitional quantifiers*, $\exists_D$ and $\forall_D$. These new quantifiers range over definitions of constants, so that we can handle constant definitions abstractly to develop semantic-aware induction heuristics that can transcend problem domains, whereas the conventional quantifiers from LiFtEr range over terms and term occurrences, so that we can handle terms and their occurrences abstractly to develop syntax based induction heuristics in a domain-agnostic style.

More specifically, each definitional quantifier takes a triple of:

– a term whose defining clauses are to be examined,
– a lambda function, which examines the relevant definitions, and
– a list of arguments, each of which is either a term or natural number. They are passed to the aforementioned lambda function to bridge the gap between the analysis of a proof goal and the analysis of relevant definitions.

For example, $\exists_D$ (const, λxs. f xs, as) returns True if λxs. f xs returns True when applied to as for *at least one* clause that defines const. Similarly, $\forall_D$ (const, λxs. f xs, as) returns True if λxs. f xs returns True when applied to as for *all* clauses that define const.

The conventional quantifiers outside and inside definitional quantifiers behave differently: inside the lambda function passed as the second argument to definitional quantifiers, conventional quantifiers' domains are based on the relevant definitions under consideration. For example, a quantifier over terms inside a definitional quantifier ranges over terms that appear in the relevant defining clause under consideration.

In the following we focus on the operational aspect of definitional quantifiers, so that readers can grasp their nature using a concrete example in Sect. 5.2.

Figure 2 illustrates the overall workflow of the SeLFiE interpreter when applied to an inductive problem and arguments of the induct tactic. In this figure, we assume that the SeLFiE assertion has only one definitional quantifier for a simpler explanation; however, in general, a SeLFiE heuristic may contain multiple definitional quantifiers. The small square, labelled as inner part, represents the lambda function passed as the second argument to this definitional quantifier, whereas outer part represents everything else in the SeLFiE assertion. Now based on this figure we explain how the SeLFiE interpreter works using the following eight steps from S1 to S8.

S1. Firstly, the SeLFiE interpreter takes a SeLFiE heuristic.

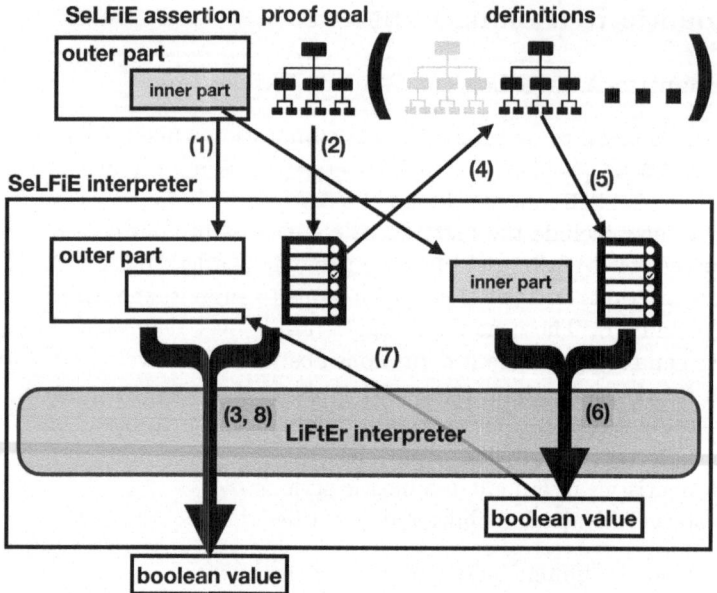

Fig. 2. The workflow of the SeLFiE interpreter.

S2. Then, the preprocessor of SeLFiE transforms the syntax tree representing the inductive problem into a look-up table. This look-up table replaces slow traversals in the syntax tree with quick accesses to term occurrences using their paths from the root node.

S3. The SeLFiE interpreter processes the outer part of the assertion using the newly implemented LiFtEr interpreter.

S4. When the SeLFiE interpreter reaches the definitional quantifier, it extracts the clauses that define the first argument of the definitional quantifier from the underlying proof context.

S5. The interpreter transforms the syntax tree representing the relevant definitions into look-up tables.

S6. The LiFtEr interpreter applies the inner part of the assertion, which is the lambda function passed as the second argument of the definitional quantifier, to the list of arguments, which is the third argument of the definitional quantifier, based on the look-up tables produced in S5.

S7. The result of S6 is then returned to the LiFtEr interpreter.

S8. The LiFtEr interpreter continues to evaluate the remaining outer part using the return value from the inner part.

We named our language SeLFiE partly because we extended LiFtEr, so that LiFtEr can call itself to support definitional quantifiers, but also because SeLFiE heuristics can attain the semantics of inductive problems using definitional quantifiers. Our motto is that:

Program 2 Syntactic analysis of more reliable generalization heuristic in SeLFiE

$\forall$ *arb_term* : term $\in$ arbitrary.
  $\exists$ *f_term* : term.
  $\exists$ *f_occ* : term_occ $\in$ *f_term*.
    $\exists$ *arb_occ* $\in$ *arb_term*.
      $\exists$ *generalize_nth* : number.
        is_nth_argument_of (*arb_occ*, *generalize_nth*, *f_occ*)
        $\wedge$
          $\exists_D$ (*f_term*, generalize_nth_argument_of, [*generalize_nth*, *f_term*])

Program 3 Definitional analysis of a generalization heuristic in SeLFiE

generalize_nth_argument_of :=
λ [*generalize_nth*, *f_term*].
  $\exists$ *lhs_occ* : term_occ. is_left_hand_side (*lhs_occ*)
  $\wedge$
    $\exists$ *nth_param_on_lhs* : term_occ.
      is_nth_argument_in (*nth_param_on_lhs*, *generalize_nth*, *lhs_occ*)
      $\wedge$
        $\exists$ *nth_param_on_rhs* : term_occ.
          $\neg$ are_of_same_term (*nth_param_on_rhs*, *nth_param_on_lhs*)
          $\wedge$
            $\exists$ *f_occ_on_rhs* : term_occ $\in$ *f_term*.
              is_nth_argument_of (*nth_param_on_rhs*, *generalize_nth*, *f_occ_on_rhs*)

We analyze inductive problems *semantically* by analyzing their relevant definitions *syntactically*.

5.2   Semantics-Aware Generalization Heuristic

We now improve the naive generalization heuristic from Sect. 4.2 in SeLFiE. More specifically, we encode the provision to the generalization heuristic discussed in Sect. 4.2 as Program 2 and Program 3. Intuitively, when applied to model_proof, these programs realise the following dialogue:

- Program 2 asks "Should we generalize *ys*, which appears as the second argument of rev2?"
- Program 3 answers "Yes, because the second argument changes from the left-hand side to the right-hand side in the second clause defining rev2."

Keeping this dialogue in mind, we examine how the SeLFiE interpreter formally processes this heuristic for our running example.

S1. We pass Program 2 and 3, and model_proof to the SeLFiE interpreter.

S2. The interpreter transforms the syntax tree representing the proof goal into a look-up table for faster processing.

S3. The SeLFiE interpreter processes the outer part for the syntax tree representing the proof goal itself. Note that the domains of quantifiers over terms and term occurrences are based on those terms and their occurrences within the proof goal itself.

In model_proof, only one variable, ys, is generalized in the arbitrary field. Therefore, for model_proof to satisfy this generalization heuristic we only have to satisfy inner existential quantifiers when *arb_term* is ys. Thus, we instantiate each existentially quantified variable in Program 2 as follows:

- *f_term* with rev2,
- *f_occ* with the sole occurrence of rev2 in the proof goal,
- *arb_occ* with the occurrence of ys on the left-hand side in the goal, and
- *generalize_nth* with 2.

Then, is_nth_argument_of returns True since ys on the left-hand side is the second argument to rev2 in the goal.

S4. When the interpreter hits $\exists_D$ with *f_term* being rev2, it extracts the two syntax trees defining rev2 from the proof context. Since $\exists_D$ is an existential quantifier, we only have to show that Program 3 returns True for one of the two equations defining rev2. In the following, we focus on the second clause, rev2 (x # xs) ys = rev2 xs (x # ys).

S5. The interpreter transforms each syntax tree representing a clause defining rev2 into a look-up table for faster processing.

S6. The interpreter evaluates Program 3 with 2 as *generalize_nth* and rev2 as *f_term*, since they are passed from Program 2. Note that the domains of quantifiers over terms and term occurrences are now all terms and term occurrences in rev2 (x # xs) ys = rev2 xs (x # ys). To satisfy Program 3 we instantiate existentially quantified variables as follows:

- *lhs_occ* with the left-hand side of the equation, rev2 (x # xs) ys,
- *nth_param_on_lhs* with the occurrence of ys, which appears as the second argument on the left-hand side,
- *f_occ_on_rhs* with the sole occurrence of rev2 on the right-hand side, and
- *nth_param_on_rhs* with the sole occurrence of x # ys, which is the second argument to rev2 bound by *f_occ_on_rhs*.

Since x # ys and ys are not the same term, the interpreter evaluates Program 3 to True for the second clause defining rev2, which is tantamount to say *we generalize the second argument of rev2 because the second argument of rev2 changes in a recursive call* in a domain-agnostic style.

S7. Program 3 returns True to Program 2.

S8. With this returned value, the interpreter evaluates Program 2 to True.

This is how Program 3 encodes the provision to the generalization heuristic discussed in Sect. 4.2. Note that the interaction between the two programs involves natural numbers, terms, and boolean values only: more complex reasoning, such as quantification over natural numbers, terms, and term occurrences, happens only within each program because each module has its own domains for

Table 1. Coincidence rates and return rates

tool	top 1	top 3	top 5	top 10
sem_ind	38.2	59.3	64.5	72.7
smart_induct	20.1	42.8	48.5	55.3

(a) Coincidence rates [%]

tool	0.2	0.5	1.0	2.0	5.0
sem_ind	8.8	24.7	47.8	69.8	86.8
smart_induct	0.0	3.5	16.9	38.3	70.2

(b) Return rates [%] within timeouts [s]

terms and term occurrences. Furthermore, it is not allowed to pass term occurrences from a syntactic analysis to a definitional analysis through definitional quantifiers. Therefore, we discuss relative locations of certain term occurrences across syntax trees, by passing natural numbers and terms from the syntax level to the definition level, as is done in this example. This clear separation between syntactic and definitional reasoning improves the readability of this heuristic.

In this particular example, we demonstrated two-level analysis of syntax trees using two SeLFiE programs. However, SeLFiE's definitional quantifiers can orchestrate reasoning on arbitrary number of levels.

6   Case Studies and Evaluations

6.1   Interactive Recommendation System

Using SeLFiE, we previously developed sem_ind, an interactive recommendation system for proof by induction in Isabelle/HOL [33]. Given an inductive problem, sem_ind produces a number of induction candidates and applies 44 SeLFiE heuristics to these candidates. Each heuristic is tagged with a certain point, representing the weight of each heuristic. Based on the sum of these points, sem_ind ranks the candidates and presents the 10 most promising ones to its users.

Nagashima evaluated sem_ind against 1,095 inductive proofs from the Archive of Formal Proofs (AFP) [21] and compared sem_ind against its predecessor, smart_induct [32], which is written in LiFtEr.

Table 1a summarizes how often sem_ind's recommendations coincide with the choices of human engineers. For example, Table 1a shows 38.2% for "sem_ind" at "top 1". This means when considering only the top one candidate recommended by sem_ind, sem_ind's recommendations coincide with the choices of human engineers for 38.2% of proof goals in the dataset. This is a 90.0% improvement compared to smart_induct, which reported 20.1 % for "top 1".

Table 1b, on the other hand, summarizes how long it takes for sem_ind to produce recommendations. For example, Table 1b shows 8.8% for "sem_ind" at "0.2". This means sem_ind managed to produce recommendations for 8.8% of proof goals in the dataset within 0.2 s of timeout. Furthermore, Nagashima also reported that the median value of the execution time of sem_ind is 1.06 s, while that of smart_induct is 2.79 s, which is a 2.63x speedup.

Program 4 Automatic inductive prover without SeLFiE

```
Auto_Solve = Thens[Auto, Solved]
PSL_WO_SeLFiE =
Ors[Auto_Solve,
    PThenOne[Dynamic (Induct), Auto_Solve]
    PThenOne[Dynamic(Induct), Thens[Auto, RepeatN(Hammer), Solved]]]
```

Program 5 Automatic inductive prover with SeLFiE

```
PSL_W_SeLFiE =
Ors[Auto_Solve,
    PThenOne[Semantic_Induct, Auto_Solve]
    PThenOne[Semantic_Induct, Thens[Auto, RepeatN(Hammer), Solved]]]
```

6.2   Automatic Proof Search Using SeLFiE

We integrated sem_ind into an automatic inductive prover written in PSL [35] and measured how SeLFiE improved PSL's automatic proof search. PSL is a domain-specific language to describe rough ideas about how to find a proof using backtracking search over tactics in Isabelle/HOL. In the following, we focus on PSL's constructs used in our evaluation leaving out irrelevant details of PSL.

Program 4 shows an example automatic inductive prover written in PSL, which we use as the baseline prover in this evaluation. The strategy is called PSL_WO_SeLFiE, and it combines three sub-strategies using the deterministic combinator Ors: it first tries the first sub-strategy, Auto_Solve, and proceeds to the second sub-strategy only if the first sub-strategy fails, and so on. Thens used in Auto_Solve is the sequential combinator, which combines Auto and Solved sequentially, and Auto in PSL corresponds to the auto tactic in Isabelle, while the following Solved checks if all sub-goals are proved by auto. Hammer represents the invocation of Sledgehammer, which is wrapped in RepeatN in Program 4. This means "repeat applying Sledgehammer to the remaining sub-goals n times where n is the number of sub-goals before applying Sledgehammer". PThenOne is the sequential parallel combinator: PThenOne takes exactly two sub-strategies and applies the second sub-strategy to the results of the first sub-strategy in parallel until at least one of them succeeds.

Dynamic (Induct) creates variants of the induct tactics with different arguments based on the given goal and combine such variants non-deterministically. However, when the interpreter produces such variants of the induct tactics using Dynamic (Induct), it does not know which one would be the most suitable induction. Therefore, the interpreter naively combines variables and arguments appearing in the proof goal to produce candidate induct tactics. In PSL, it is the subsequent sub-strategies that are to identify the right arguments for the induct tactic: PThenOne [Dynamic (Induct), Auto_Solve], for example, keeps applying auto to sub-goals emerging after applying the induct tactic with various sequences of arguments until it finds a sequence that results in sub-goals that are all proved by `auto`.

The drawback of this approach is that PSL's interpreter cannot identify the appropriate arguments for the `induct` tactic if it cannot complete a proof search: for difficult inductive problems, the interpreter often fails to complete a proof search within a realistic timeout because `Dynamic (Induct)` tends to produce a large number of induction candidates and the necessary proof steps after applying the `induct` tactic tend to be complicated. What was lacking was the mechanism to identify promising induction candidates without relying on a proof search, so that PSL's interpreter can spend limited computational resources for a small number of promising candidates to complete a proof search. For this reason, we integrated `sem_ind` into PSL, and we counted how many goals are proved within each timeout.

Program 5 shows the new automatic prover. Here, `Semantic_Induct` represents `sem_ind` integrated into PSL's environment. We highlighted the differences in Program 5 from Program 4 to clarify that we are using almost the same PSL strategy for a fair comparison except for the introduction of `Semantic_Induct`.

For our evaluation, we used 12 Isabelle theory files from 8 projects about various topics in the AFP, which in total include 347 proofs by induction. These projects are about the depth-first search [40], binomial heaps [25], a boolean expression checker [37], multi-dimensional binary search trees [42], the priority search tree [23], linear temporal logic [45], imperative programming language Simpl [44], and program verification competition [24]. We conducted this evaluation on a MacBook Pro (15-in., 2019) with 2.6 GHz Intel Core i7 6-core memory 32 GB 2400 MHz MHz DDR4, and the reported execution times are based on elapsed real time.

Table 2. Success rates and speedup

timeouts	Program 5	Program 4		speedup [times]	occurrence
0.3[s]	11.0%	1.2%		x < 1.0	3 (2.4%)
1.0[s]	25.6%	1.7%		1.0 ≤ x < 5.0	64 (50.8%)
3.0[s]	28.2%	21.9%		5.0 ≤ x < 10.0	44 (34.9%)
10.0[s]	34.9%	28.0%		10.0 ≤ x < 15.0	9 (7.1%)
30.0[s]	45.8%	38.3%		15.0 ≤ x <	6 (4.8%)
(a) Success rates				(b) Speedup of execution time	

Table 2a shows how many inductive problems were proved by each program within each timeout. For example, the timeout of 0.3[s] for Program 5 has 11.0%. This means Program 5 proved 11.0% inductive problems in the dataset within 0.3 s. For a fair comparison we included not only the time spent by tactics for proof search but also the time spent by `sem_ind` when measuring the execution time of each proof search. As shown in Table 2a, PSL enhanced with `sem_ind` proved more inductive problems than PSL without sem_ind for various timeouts. For 30.0 s of timeout, PSL with sem_ind proved 159 inductive problems, while PSL without sem_ind proved 133 problems only. 126 problems were proved by both provers within this timeout. For each problem proved by both programs within 30.0 s, we computed the speedup of execution time spent to complete each proof search. For example, Program 5 spent 0.325 s and Program 4 spent 2.171 s to prove a lemma named nexts_set in DFS.thy. Therefore, the speedup of execution time for this lemma is (2.171 / 0.325) = 6.68.

Table 2b shows the distribution of speedup observed among such problems. For example, the second row reads $1.0 \leq x < 5.0$ and 64.0 (50.8%), and this means that Program 5 achieved between 1.0x to 5.0x speedup compared to Program 4 for 64 inductive problems proved by both provers. As shown in this table, we confirmed that Program 5 achieved speedups over Program 4 except for 3 cases, which constitutes 2.4% of problems proved by both provers within 30.0 s of timeout. The median value for speedup is 4.48x.

7   Conclusion

We presented SeLFiE, a boolean-query language to encode induction heuristics. The abstraction brought by definitional quantifiers allow SeLFiE to transcend problem domains while analysing not only the syntactic structures of inductive problems but also definitions of relevant constants in a modular style.

Our conservative extension to LiFtEr's syntax allows us to take advantage of LiFtEr's domain-agnosticism, while adding the capability to reason on the semantics of proof goals. To realise such extension, we implemented SeLFiE's interpreter from scratch: since LiFtEr's original interpreter was not designed with definitional reasoning in mind, it did not support even lambda abstraction or function application, and suffered from poor performance, incremental improvement was not realistic.

Nagashima implemented sem_ind in SeLFiE, and we integrated sem_ind into PSL and built an automatic inductive prover. Our experiment showed that compared to the baseline prover our inductive prover based on SeLFiE achieves $1.4 \cdot 10^3\%$ improvement of success rate for 1.0 s of timeout as well as a 4.48x speedup as the median value.

The final goal of this project is to build a strong inductive prover. It remains our future work to further strengthen the automatic prover introduced in Sect. 6, by incorporating two conjecturing mechanisms, top-down conjecturing [36] and bottom-up conjecturing [17], into our system.

8   Related Work

A well-known approach for inductive theorem proving is the Boyer-Moore waterfall model [26], which was invented for a first-order logic on Common Lisp [18]. In the original waterfall model, a prover tries to apply any of the six techniques, including simplification, generalization and induction. If any of these techniques works, the prover stores the resulting sub-goals in a pool and continues to apply the techniques until it empties the pool.

ACL2 [27] is the latest incarnation of this line of work with industrial applications [20]. To decide how to apply induction, ACL2 estimates how good each induction scheme is by computing a score, called *hitting ratio*, based on a fixed formula [4, 28], and it proceeds with the induction scheme with the highest hitting ratio. Heras *et al.* used ML4PG learning method to find patterns to generalize and transfer inductive proofs from one domain to another in ACL2 [14]. Instead of computing a hitting ratio, we provide SeLFiE as a language, so that Isabelle experts can encode their expertise as assertions.

There are ongoing attempts to extend saturation-based superposition provers with induction: Cruanes presented an extension of typed superposition that can perform structural induction [8], while Reger *et al.* incorporated lightweight automated induction [43] to the Vampire prover [22] and Hajdú *et al.* extended it to cover induction with generalization [15]. Contrary to their work, our approach to proof by induction uses Isabelle's default induct tactic, which we can use for arbitrary data types.

For more expressive logics, Jiang *et al.* employed multiple waterfalls [16] in HOL Light [13]. However, to decide induction variables, they naively picked the first free variable with recursive type and left the selection of promising induction variables as future work. Passmore *et al.* developed the Imandra automated reasoning system [41], which also uses the waterfall model for its typed higher-order setting. For Isabelle/HOL, Dixon *et al.* developed IsaPlanner [11], a generic framework to encode proof plans [5]. IsaPlanner can incorporate reasoning techniques, such as rippling [7], for proof by induction. For generalization, however, IsaPlanner naively generalizes all non-induction variables [10].

Machine learning tools for tactic-based theorem proving mainly focus on tactic recommendations and premise selections, leaving the problem of arguments selection for tactics as an open question when arguments are terms [1, 12, 34]. Instead of relying on machine learning algorithms, we developed a language, in which one can explicitly encode heuristics. We plan to use SeLFiE as a feature extractor for machine learning algorithms: by applying SeLFiE heuristics to inductive problems, we can convert each pair of an inductive problem and induction arguments to an array of boolean values, which is amenable for machine learning algorithms. The application of SeLFiE as a preprocessor for machine learning algorithms remains as our future work.

Acknowledgement. We thank the anonymous reviewers for the useful feedback, both at Tests and Proofs 2022 and other conferences. This work was supported by the following grants:

– NII under NII-Internship Program 2019-2nd call,

– the European Regional Development Fund under the project AI & Reasoning. (reg.no.CZ.02.1.01/0.0/0.0/15_003/0000466)

References

1. Blaauwbroek, L., Urban, J., Geuvers, H.: Tactic learning and proving for the Coq proof assistant. In: LPAR 2020: 23rd International Conference on Logic for Programming, Artificial Intelligence and Reasoning, Alicante, Spain (2020)
2. Blanchette, J.C., Böhme, S., Paulson, L.C.: Extending sledgehammer with SMT solvers. In: Bjørner, N., Sofronie-Stokkermans, V. (eds.) CADE 2011. LNCS (LNAI), vol. 6803, pp. 116–130. Springer, Heidelberg (2011). https://doi.org/10.1007/978-3-642-22438-6_11
3. Giesl, J., Hähnle, R. (eds.): IJCAR 2010. LNCS (LNAI), vol. 6173. Springer, Heidelberg (2010). https://doi.org/10.1007/978-3-642-14203-1
4. Boyer, R.S., Moore, J.S.: A Computational Logic Handbook, Perspectives in Computing, vol. 23. Academic Press (1979)
5. Bundy, A.: The use of explicit plans to guide inductive proofs. In: Lusk, E., Overbeek, R. (eds.) CADE 1988. LNCS, vol. 310, pp. 111–120. Springer, Heidelberg (1988). https://doi.org/10.1007/BFb0012826
6. Bundy, A.: The automation of proof by mathematical induction. In: Robinson, J.A., Voronkov, A. (eds.) Handbook of Automated Reasoning (in 2 volumes), pp. 845–911. Elsevier and MIT Press (2001)
7. Bundy, A., Stevens, A., van Harmelen, F., Ireland, A., Smaill, A.: Rippling: A heuristic for guiding inductive proofs. Artif. Intell. **62**, 185–253 (1993)
8. Cruanes, S.: Superposition with structural induction. In: Dixon, C., Finger, M. (eds.) FroCoS 2017. LNCS (LNAI), vol. 10483, pp. 172–188. Springer, Cham (2017). https://doi.org/10.1007/978-3-319-66167-4_10
9. Czajka, Ł, Kaliszyk, C.: Hammer for Coq: automation for dependent type theory. J. Autom. Reasoning, 423–453 (2018). https://doi.org/10.1007/s10817-018-9458-4
10. Dixon, L.: A proof planning framework for Isabelle. Ph.D. thesis, University of Edinburgh, UK (2006). http://hdl.handle.net/1842/1250
11. Dixon, L., Fleuriot, J.: IsaPlanner: a prototype proof planner in Isabelle. In: Baader, F. (ed.) CADE 2003. LNCS (LNAI), vol. 2741, pp. 279–283. Springer, Heidelberg (2003). https://doi.org/10.1007/978-3-540-45085-6_22
12. Gauthier, T., Kaliszyk, C., Urban, J.: TacticToe: learning to reason with HOL4 tactics. In: LPAR-21, 21st International Conference on Logic for Programming, Artificial Intelligence and Reasoning, Maun, Botswana, 7–12 May 2017 (2017)
13. Harrison, J.: HOL light: a tutorial introduction. In: Srivas, M., Camilleri, A. (eds.) FMCAD 1996. LNCS, vol. 1166, pp. 265–269. Springer, Heidelberg (1996). https://doi.org/10.1007/BFb0031814
14. Heras, J., Komendantskaya, E., Johansson, M., Maclean, E.: Proof-pattern recognition and lemma discovery in ACL2. In: Logic for Programming, Artificial Intelligence, and Reasoning - 19th International Conference, LPAR-19, Stellenbosch, South Africa, 14–19 December 2013. Proceedings (2013)
15. Hozzová, P., Kovács, L., Schoisswohl, J., Voronkov, A.: Induction with generalization in superposition reasoning. EasyChair Preprint no. 2468 (EasyChair 2020) (2020)
16. Jiang, Y., Papapanagiotou, P., Fleuriot, J.: Machine learning for inductive theorem proving. In: Fleuriot, J., Wang, D., Calmet, J. (eds.) AISC 2018. LNCS (LNAI), vol. 11110, pp. 87–103. Springer, Cham (2018). https://doi.org/10.1007/978-3-319-99957-9_6
17. Watt, S.M., Davenport, J.H., Sexton, A.P., Sojka, P., Urban, J. (eds.): CICM 2014. LNCS (LNAI), vol. 8543. Springer, Cham (2014). https://doi.org/10.1007/978-3-319-08434-3

18. Steele Jr., G.L.: An overview of common Lisp. In: Proceedings of the 1982 ACM Symposium on LISP and Functional Programming, LFP 1980, 15–18 August 1982, Pittsburgh, PA, USA (1982)
19. Kaliszyk, C., Urban, J.: Hol(y)hammer: online ATP service for HOL light. Math. Comput. Sci. **9**, 5–22 (2015)
20. Kaufmann, M., Moore, J.S.: An industrial strength theorem prover for a logic based on common lisp. IEEE Trans. Software Eng. **23**, 203–213 (1997)
21. Klein, G., Nipkow, T., Paulson, L., Thiemann, R.: The Archive of Formal Proofs (2004). https://www.isa-afp.org/
22. Kovács, L., Voronkov, A.: First-order theorem proving and VAMPIRE. In: Sharygina, N., Veith, H. (eds.) CAV 2013. LNCS, vol. 8044, pp. 1–35. Springer, Heidelberg (2013). https://doi.org/10.1007/978-3-642-39799-8_1
23. Lammich, P., Nipkow, T.: Priority search trees. Arch. Formal Proofs (2019)
24. Lammich, P., Wimmer, S.: Verifythis 2019 - polished Isabelle solutions. Arch. Formal Proofs (2019)
25. Meis, R., Nielsen, F., Lammich, P.: Binomial heaps and skew binomial heaps. Arch. Formal Proofs (2010)
26. Moore, J.S.: Computational logic: structure sharing and proof of program properties. Ph.D. thesis, University of Edinburgh, UK (1973)
27. Moore, J.S.: Symbolic simulation: an ACL2 approach. In: Formal Methods in Computer-Aided Design, Second International Conference, FMCAD 1998, Palo Alto, California, USA, 4–6 November 1998, Proceedings (1998)
28. Moore, J.S., Wirth, C.: Automation of mathematical induction as part of the history of logic. CoRR abs/1309.6226 (2013). http://arxiv.org/abs/1309.6226
29. de Moura, L.M., Kong, S., Avigad, J., van Doorn, F., von Raumer, J.: The lean theorem prover (system description). In: Automated Deduction - CADE-25 - 25th International Conference on Automated Deduction, Berlin, Germany (2015)
30. Nagashima, Y.: Data61/PSL (2017). https://github.com/data61/PSL/releases/tag/v0.2.1-alpha
31. Nagashima, Y.: LiFtEr: language to encode induction heuristics for Isabelle/HOL. In: Lin, A.W. (ed.) APLAS 2019. LNCS, vol. 11893, pp. 266–287. Springer, Cham (2019). https://doi.org/10.1007/978-3-030-34175-6_14
32. Nagashima, Y.: Smart induction for Isabelle/HOL (tool paper). In: Proceedings of the 20th Conference on Formal Methods in Computer-Aided Design - FMCAD 2020 (2020)
33. Nagashima, Y.: Faster smarter proof by induction in Isabelle/HOL. In: Zhou, Z. (ed.) Proceedings of the Thirtieth International Joint Conference on Artificial Intelligence, IJCAI 2021, Virtual Event/Montreal, Canada, 19–27 August 2021, pp. 1981–1988. ijcai.org (2021). https://doi.org/10.24963/ijcai.2021/273
34. Nagashima, Y., He, Y.: PaMpeR: proof method recommendation system for Isabelle/HOL. In: Proceedings of the 33rd ACM/IEEE International Conference on Automated Software Engineering, ASE 2018, Montpellier, France (2018)
35. Nagashima, Y., Kumar, R.: A proof strategy language and proof script generation for Isabelle/HOL. In: de Moura, L. (ed.) Automated Deduction - CADE 26–26th International Conference on Automated Deduction. Gothenburg, Sweden (2017)
36. Nagashima, Y., Parsert, J.: Goal-oriented conjecturing for Isabelle/HOL. In: Rabe, F., Farmer, W.M., Passmore, G.O., Youssef, A. (eds.) CICM 2018. LNCS (LNAI), vol. 11006, pp. 225–231. Springer, Cham (2018). https://doi.org/10.1007/978-3-319-96812-4_19
37. Nipkow, T.: Boolean expression checkers. Arch. Formal Proofs (2014)

38. Nipkow, T., Klein, G.: Concrete Semantics. Springer, Cham (2014). https://doi. org/10.1007/978-3-319-10542-0
39. Nipkow, T., Wenzel, M., Paulson, L.C. (eds.): Isabelle/HOL. LNCS, vol. 2283. Springer, Heidelberg (2002). https://doi.org/10.1007/3-540-45949-9
40. Nishihara, T., Minamide, Y.: Depth first search. Arch. Formal Proofs (2004)
41. Passmore, G., et al.: The Imandra automated reasoning system (system description). In: Peltier, N., Sofronie-Stokkermans, V. (eds.) IJCAR 2020. LNCS (LNAI), vol. 12167, pp. 464–471. Springer, Cham (2020). https://doi.org/10.1007/978-3-030-51054-1_30
42. Rau, M.: Multidimensional binary search trees. Arch. Formal Proofs (2019)
43. Reger, G., Voronkov, A.: Induction in saturation-based proof search. In: Fontaine, P. (ed.) CADE 2019. LNCS (LNAI), vol. 11716, pp. 477–494. Springer, Cham (2019). https://doi.org/10.1007/978-3-030-29436-6_28
44. Schirmer, N.: A sequential imperative programming language syntax, semantics, Hoare logics and verification environment. Arch. Formal Proofs (2008)
45. Sickert, S.: Linear temporal logic. Arch. Formal Proofs (2016)
46. Slind, K., Norrish, M.: A brief overview of HOL4. In: Mohamed, O.A., Muñoz, C., Tahar, S. (eds.) TPHOLs 2008. LNCS, vol. 5170, pp. 28–32. Springer, Heidelberg (2008). https://doi.org/10.1007/978-3-540-71067-7_6

Effective Testing

Effective testing

Fuzzing and Delta Debugging And-Inverter Graph Verification Tools

Daniela Kaufmann[1]([✉])[iD] and Armin Biere[2][iD]

[1] Johannes Kepler University Linz & Software Competence Center, Hagenberg, Austria
daniela.kaufmann@scch.at
[2] Albert-Ludwigs-University Freiburg, Freiburg, Germany
biere@cs.uni-freiburg.de

Abstract. Ensuring correctness of verification tools is equally important as the correctness of the actual problems they try to establish. In this paper we evaluate automated fuzzing and debugging techniques applied to state-of-the-art multiplier verification tools, which take the common gate-level representation of and-inverter graphs as input. With a generation- and mutation-based fuzzing approach our tools are able to uncover major faults in verification tools including crashes as well as inaccurate verification results. Additionally we demonstrate the usefulness of certificates for automated reasoning tools. We further apply delta debugging techniques and show their effectiveness in reducing failure-inducing inputs.

1 Introduction

Hardware verification and particularly formal verification of arithmetic circuits is important to prevent issues like the infamous Pentium FDIV bug [31]. In recent years verification of gate-level integer multipliers has made significant progress and it has been shown that algebraic reasoning techniques are particularly successful [6,14,20,21]. In this line of work the given multiplier circuit is modeled as a set of polynomials that generates a Gröbner basis. Using a polynomial reduction algorithm, it is checked whether the specification, also modeled as a polynomial, is implied by the circuit polynomials. We refer the reader to [11] for a more detailed introduction to circuit verification using computer algebra.

Sophisticated reduction engines geared towards the automatic verification of multipliers are implemented in the two state-of-the-art tools DYPOSUB [21] and AMULET2 [13]. These tools receive multiplier circuits, given as and-inverter graphs (AIGs) [18], internally generate a polynomial encoding and perform the

This work is supported by the LIT AI Lab funded by the State of Upper Austria, by the Federal Ministry for Climate Action, Environment, Energy, Mobility, Innovation and Technology (BMK), the Federal Ministry for Digital and Economic Affairs (BMDW), and the Province of Upper Austria in the frame of the COMET–Competence Centers for Excellent Technologies Programme and the COMET Module DEPS managed by the Austrian Research Promotion Agency FFG.

L. Kovács and K. Meinke (Eds.): TAP 2022, LNCS 13361, pp. 69–88, 2022.
https://doi.org/10.1007/978-3-031-09827-7_5

reduction. DᴙPᴏSᴜʙ simply provides a yes-or-no answer on the correctness of the circuit. Our tool AMᴜʟᴇᴛ2 additionally provides a proof certificate for correct circuits, and counterexamples in the case of an incorrect circuit.

Proofs and concrete counterexamples (models) provide an additional layer of confidence for the yes-or-no answer of the verification tool. Especially, as we will see in our experimental evaluation, DʏPᴏSᴜʙ and AMᴜʟᴇᴛ2 do not always agree on the verification results. Using only the yes-or-no answer it is hard to decide which tool is correct on disagreement. In those cases counterexamples and particularly proof certificates are not only essential to determine correctness, but they can also be used to debug correctness errors. Clearly, robustness and correctness are essential for the usability of tools. An incorrect verifier might for instance fail to detect errors in the implementation of a multiplier, i.e., reporting that the circuit is correct although it contains a bug.

Current research on algebraic multiplier verification focuses on efficiency and automation. However, we are not aware of research in the direction of automated fuzzing and debugging techniques for these tools that use AIGs as input. In automated reasoning fuzzing and delta debugging techniques have originally been applied for satisfiability modulo theories (SMT) [4], satisfiabilty (SAT) and quantified Boolean formulas (QBF) solvers [5], and interactive provers [19] and continue to be an essential part of the solver-developer tool-box [1,3,17,23,27, 28,30]. For a general introduction to fuzzing we refer to the Fuzzing Book [35].

This paper gives an incentive towards investing effort in automated testing and debugging of automated reasoning tools, with the focus on hardware verification. We present our generation-based fuzzer MᴜʟᴛAIGᴇɴFᴜᴢᴢᴇʀ that produces valid multiplier circuits by randomly arranging modules of multiplier circuits. The goal of these benchmarks is to avoid overfitting of automated reasoning tools to those benchmark families that are currently publicly available and used, e.g., benchmarks generated by AMG [9], GᴇɴMᴜʟ [22], and Mᴜʟᴛ-ɢᴇɴ [32].

Additionally we present a mutation-based fuzzing tool AIGᴏFᴜᴢᴢɪɴɢ that applies small modifications to a given AIG, e.g., flipping signs. These mutations may or may not affect the correctness of the AIG. AIGᴏFᴜᴢᴢɪɴɢ is not tailored towards multiplier verification and may be applied to any given AIG, hence we consider it a general purpose mutation-based AIG fuzzer.

We expect that some of the fuzzing techniques deliver failure-inducing benchmarks that lead to solver crashes or incorrect results. Debugging failure-inducing benchmarks and locating the program error is typically a very time consuming task. Usually only a small part of the AIG raises the failing behavior, whereas the remainder of the graph is redundant with respect to the error. Reduction of tree-structured inputs is for example also discussed in [8]. In this paper we present a delta debugging tool AIGᴅᴅ2 that allows to shrink failure-inducing inputs while preserving the failures. AIGᴅᴅ2 is an updated re-implementation of AIGᴅᴅ [2]; it still provides the same functionality as AIGᴅᴅ, but has additional options to keep a multiplier-alike input/output signature of the AIG.

We evaluate the presented tools MULTAIGENFUZZER, AIGOFUZZING, and AIGDD2 on the state-of-the-art multiplier verification tools DYPOSUB [21] and AMULET2 [13]. Our experiments show that both tools have issues. For instance, we got benchmarks where DYPOSUB miss-classifies the correctness of multipliers, i.e., claims that a multiplier is correct although it calculates $3 \cdot 3 = 13$. Our tool AMULET2 did not produce wrong results, however it has several robustness issues. The presented approach allowed us to fix these issues in AMULET2.

The remainder of the paper is structured as follows, Sect. 2 discusses the input format AIG and introduces the used fuzzing and debugging techniques. In Sect. 3, 4 and 5 we present our tools MULTAIGENFUZZER, AIGOFUZZING, and AIGDD2. Section 6 discusses automated testing followed by our experiments in Sect. 7, before we conclude in Sect. 8 with a summary of learned lessons.

2 Fuzzing, Delta Debugging and AIGs

An *And-Inverter Graph* (AIG) [18] is a common representation of gate-level hardware circuits that is very efficient to handle and manipulate. An AIG is a directed acyclic graph that consists only of two-input nodes that represent logical conjunction. The edges may contain markings that indicate logical negations, i.e., inverters. The constant 0 in an AIG represents FALSE, and 1 represents TRUE. A small sample AIG is shown in Fig. 2.

Fuzzing is a technique for automated software testing. The original idea was to treat the program as a black-box and use random, invalid and unexpected inputs to detect failures and tool crashes, e.g., buffer overflows.

The origin of fuzz testing goes back to the 90's, where it was demonstrated that random inputs are able to detect many errors in UNIX command line programs [26]. Since then a variety of automated testing approaches and tools have been developed, such as CLUSTERFUZZ by Google, or ONEFUZZ by Microsoft. The original testing approaches have recently been repeated on current UNIX systems [25] and still derived failure rates between 12%–19%. Especially pointer and array errors still seem to be present in state-of-the-art utilities.

Fuzzing approaches can be distinguished depending on whether they use existing input [35] or not. If a fuzzer reuses existing input, it is called *mutation-based* fuzzer, as it mutates the provided input by making small modifications. In contrast *generation-based* fuzzers generate input from scratch and hence do not depend on the existence of inputs (also called *seeds*).

A second differentiation relates to whether the fuzzer makes use of the structure of the tested program. A *white-box* fuzzer uses program analysis, e.g., symbolic execution [7] of either source-code or binaries, to systematically generate inputs with the goal to increase code coverage of the test program executing these inputs. On the opposite range of the spectrum, *black-box* fuzzing is completely unaware of the internal structure of the program under test. Hence, testing using black-box fuzzers is typically faster than white-box fuzzing and can easily be parallelized. However, black-box fuzzers may only trigger easy-to-reach bugs without ever reaching certain potentially harmful corner cases.

In this paper, we discuss two fuzzing tools. First MULTAIGENFUZZER, a generation-based black-box fuzzer, generating multiplier circuits from scratch. Our second fuzzing tool AIGoFUZZING is a mutation-based black-box fuzzer. It has to be provided with an input seed, which is mutated without violating structural constraints, i.e., violating the input/output signature of multipliers.

Delta debugging aims to reduce the manual workload of debugging software problems by minimizing failure-inducing inputs [33,34,36]. In a nutshell, delta debugging uses a (binary) search strategy to repeatedly remove smaller and smaller parts of the failure-inducing input until a minimal fix point is reached that triggers the same failure in the program. That is, our delta debugger AIGDD2 removes nodes from AIGs to narrow down the failure cause.

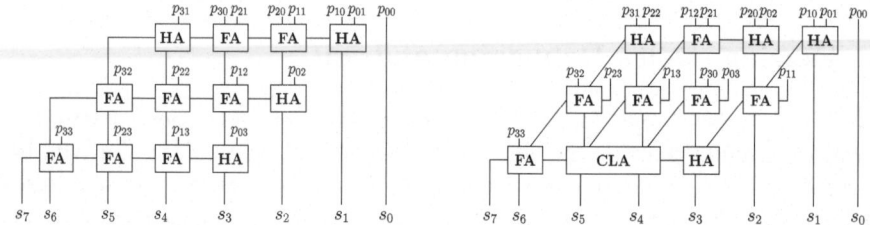

Fig. 1. 4-bit multipliers generated by MULTAIGENFUZZER, where $p_{ij} = a_i b_j$. The left multiplier is generated using only full and half adders, the right multiplier uses a carry-lookahead module (CLA).

3   Generation-based Fuzzing for Multipliers

We consider n-bit multiplier circuits, with $2n$ inputs $a_{n-1} \ldots a_0, b_{n-1} \ldots b_0$ and $2n$ outputs $s_{2n-1} \ldots s_0$, which compute $\sum_{i=0}^{2n-1} 2^i s_i = \left(\sum_{i=0}^{n-1} 2^i a_i\right)\left(\sum_{i=0}^{n-1} 2^i b_i\right)$.

In general multipliers are made of three components: (i) partial product generation (PPG), where the n^2 partial products $p_{ij} = a_i b_j$, $0 \leq i, j \leq n-1$ are generated; (ii) partial product accumulation (PPA), where the partial products are summed up to two layers, and (iii) a final-stage adder (FSA) [29].

For each component several hardware algorithms are available, which can be combined to derive a correct multiplier architecture, e.g., using a Booth-encoding for PPG, a Wallace-tree as PPA and a carry-lookahead adder as FSA. For example, the AMG [9] tool provides 22 components that can be combined to generate 192 different multiplier circuits (of arbitrary bit-width).

All these components have certain patterns and their number is limited. Thus hardware verification tools tend to be biased towards these architectures. For instance, our tool AMULET2 substitutes complex FSAs, such as carry-lookahead adders, with an equivalent ripple-carry adder. The correctness of the replacement is checked using SAT solving and the rewritten circuit is verified using computer algebra [14]. To apply adder substitution we implemented a heuristic algorithm that uses syntactic pattern matching to identify the FSA.

The purpose of our novel generation-based fuzzing tool MULTAIGENFUZZER is to generate correct multipliers, where the individual components do not follow a specific pattern to avoid overfitting solvers to available benchmarks. Figure 1 shows abstract examples for multipliers generated by MULTAIGENFUZZER.

We follow the natural approach of long multiplication: first all partial products are generated and uniquely assigned to $2n$ column-wise slices $S_0, \ldots, S_{2n-1}$, i.e., the partial product p_{ij} with $0 \leq i, j \leq n-1$ belongs to slice S_{i+j}. All partial products in the slices are added together to receive the output of the multiplication. Carries are propagated to the next bigger slices. Since addition is commutative the order of the partial products within a slice does not matter.

In MULTAIGENFUZZER we generate all n^2 partial products p_{ij} using single and-gates. The integration of Booth encoding is part of future work. After the partial products are assigned to $2n$ slices, we repeatedly select two or three random elements of a randomly picked slice S_k and add them together using a half or full adder. The sum output bit of the half or full adder will be added to S_k, the carry will be added to S_{k+1}. Our tool supports two models with different density for half and full adders, which are selected randomly. We repeat these steps until all slices contain at maximum two elements.

In the last step we generate a random FSA using a mixture of full and half adders and carry-lookahead addition. We traverse from the smallest slice S_0 to the largest slice S_{2n-1} and instantiate adder modules appropriate for the number of elements in a slice S_k. If the considered slice contains only a single element, this element becomes the k-th output of the multiplier. If a slice contains two elements we use a half adder to sum up the elements. The sum output is the k-th output bit of the multiplier circuit, the carry is inserted into the next slice S_{k+1}. In the case a slice S_k contains three elements, e.g., two elements from the partial product accumulation and one carry from the FSA module of slice S_{k-1}, we further randomly choose between a full adder or a carry-lookahead adder. We randomly select the carry-lookahead module with a probability of $2/3$.

If a full adder is selected, summing up the three elements follows the same procedure as for a half adder. If a carry-lookahead adder is chosen, we first count the number of coherent slices $S_{k+1}, \ldots, S_{2n-1}$ that contain two elements, as this will be the maximum size max of the carry-lookahead adder. We select a random number in $[1, max]$ to choose the actual size of the carry-lookahead adder.

The sum outputs of a carry-lookahead adder are generated using XOR gates. The carries can be computed recursively or iteratively. We have decided to use three modes in MULTAIGENFUZZER, which are selected randomly. In the first mode all carries of the carry lookahead adder are computed recursively. In a second mode all carries are computed iteratively. In the third mode we mix between the first two options, i.e., we randomly select for each carry whether it is generated recursively or iteratively.

4   Mutation-based Fuzzing for AIGs

In contrast to MULTAIGENFUZZER, presented in the previous section, our second fuzzer AIGoFUZZING is a mutation-based fuzzing tool and requires existing seeds. The purpose of AIGoFUZZING is to make small mutations in a given AIG that may or may not change the specification of the graph. Our fuzzer AIGoFUZZING is not specifically designed for multiplier verification and can be used on any given general-purpose AIG.

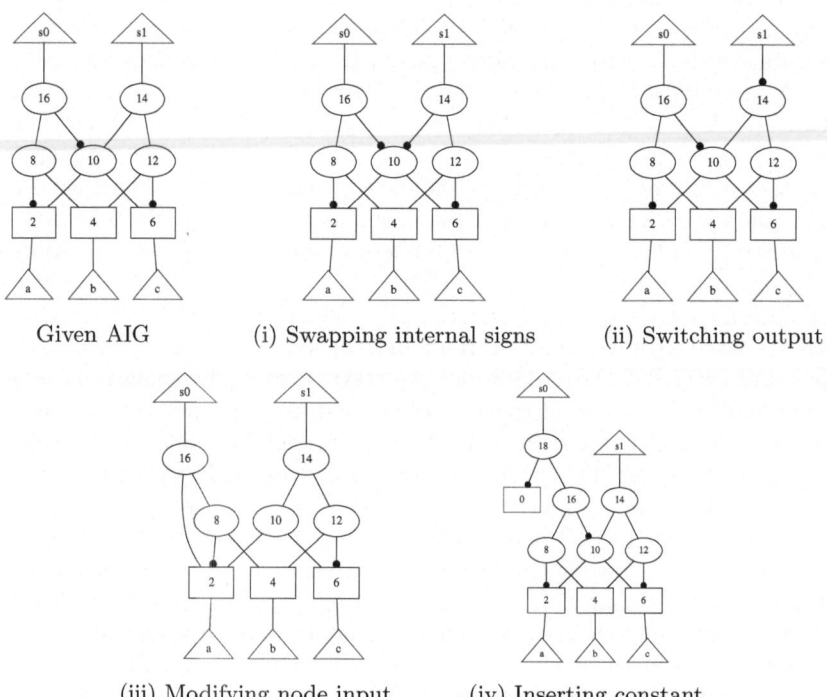

Given AIG   (i) Swapping internal signs   (ii) Switching output

(iii) Modifying node input   (iv) Inserting constant

Fig. 2. A given AIG, that is mutated using one of the four available muations (i)–(iv)

We currently support four classes of mutations, which can be selected optionally. The provided mutations are (i) swapping internal signs, (ii) switching an output signal, (iii) modifying the input of a node, and (iv) inserting constants 0 and 1. An example is given in Fig. 2. The location of the mutation is chosen randomly and if not specified otherwise AIGoFUZZING only executes single mutations. Applying multiple modifications is possible via the options.

Swapping internal signs may happen in two ways. Either the sign of a single randomly chosen edge is flipped, or a complete node is flipped by switching the signs of all outgoing edges. Switching an output signal is actually a special case of flipping a complete node, yet we have decided to include this as a separate mutation technique, as AIGs that encode miters consist only of one output bit. Modifying the sign of the output bit allows us to negate the specification of the whole AIG. In all cases, flipping the signs usually changes the behavior of the input AIG, i.e., in our setting correct multipliers will become incorrect. We will revisit this claim later when discussing Table 3 in the experimental evaluation.

In the third mutation technique we modify the input of an AIG node. We randomly select a parent node, one of its children that is replaced, and a topologically smaller node that will become the new input. It is important that we choose a new child node that is topologically smaller than the parent to prevent the formation of cycles in the generated AIG. Of course we do expect that most of the time modifying the input of an AIG affects the correctness of the AIG.

The fourth mutation technique injects constants 0 and 1 into the AIG. First we randomly select which constant is inserted and then whether it will be added as an input of a random node using logical conjunction or disjunction. This yields four possible combinations (0, conjunction), (0, disjunction), (1, conjunction), and (1, disjunction). Since $v \land 1 = v \lor 0 = v$ for any node v, integrating the constant 0 using disjunction, and 1 using conjunction will not affect the correctness of the AIG. On the other hand $v \land 0 = 0$ and $v \lor 1 = 1$. Thus the combinations (0, conjunction) and (1, disjunction) will change the behavior. This can be seen in Table 3 in the experimental evaluation that we have a 50:50 ratio on the correctness of the tested multipliers.

5   Delta Debugging with Slices

AIGDD2 is a delta debugging tool that is designed to minimize failure-inducing AIGs, while preserving the errors. AIGDD2 is a re-implementation of AIGDD [2], which has been developed for model checking purposes, whereas this paper focuses on multiplier verification tools. In particular we extend the functionality of AIGDD by adding options that limit structural changes to the AIG, e.g., to maintain the signature of multiplier circuits. Furthermore we include a slicing based delta debugging approach that allows us to shrink the bit-width of multipliers.

The goal of AIGDD2 is to automatically reduce the size of failure-inducing AIGs until a fixed point is reached. AIGDD2 reads and executes the failure-inducing AIG on the buggy program. The exit code of the faulty program is stored and will be used as golden reference value.

First AIGDD2 slices the given AIG by removing the most significant inputs and outputs. We set these bits to 0 and propagate the constant, e.g., for node v with inputs w and 0, we deduce $w \land 0 = v = 0$. The precise number of inputs and outputs that are set to 0 in each iteration can be specified via the options, e.g., for shrinking multipliers we always remove two outputs and two inputs. We repeat slicing as long as the shrunken AIG triggers the failing behavior.

Then AIGDD2 uses a binary-search based approach to further shrink the size of the sliced AIG. We start by setting the first half of AIG nodes to the constant 0, which is propagated. We then test whether the resulting AIG still triggers the failing behavior in the buggy program. If so, we proceed with the second half of the AIG nodes. If the buggy program returns a different exit code, we try again with setting the first half of AIG nodes to the constant 1. If setting the first half of AIG nodes to a constant 0 or 1 does not trigger the failure of the tested program, we split the first half of the AIG nodes into two parts and repeat the steps above. We repeatedly narrow down the search space until we have set individual AIG nodes to 0 resp. 1.

At this point we want to emphasize that it is not guaranteed that AIGDD2 generates the smallest possible failure-inducing input. However, while debugging errors it is most of the time not needed to generate a minimal example. The desire is to generate small inspectable inputs that can be used for efficient debugging.

6   Fuzzing, Tests and Proofs

We evaluate our presented fuzzing and debugging approaches on the most recent multiplier verification tools DyPoSub [21] and AMulet2 [13], where we are the authors of the latter. Both tools apply algebraic reasoning to decide on the correctness of the given multiplier. The specification of the circuit, modeled as a polynomial is reduced by the polynomial representation induced by the AIG with regard to a reverse topological variable ordering. The circuit is correct if and only if the final result is zero.

In our tool AMulet2 we replace complex FSAs, which are a bottleneck for algebraic reasoning [14], with a simple ripple-carry adder that is easy for algebraic reasoning. The rewritten circuit is verified in AMulet2 using a static variable ordering. We use version AMulet2.1 [10] for our experimental evaluation.

The tool DyPoSub applies a dynamic variable ordering attempt, i.e., after each reduction step the change in the size of the current intermediate reduction result is investigated. If the growth is above a certain threshold, the reduction step is undone and a different variable is considered for reduction. This approach aims to guarantee that the intermediate reduction results do not explode.

We use our presented fuzzing and delta debugging techniques in two ways. First, we investigate the *robustness* of the verification tools. More precisely we generate multiplier circuits using MultAIGenFuzzer that will be mutated using AIGoFuzzing. We run these benchmarks on AMulet2.1 and DyPoSub to detect critical software failures and crashes. With these results we fixed several issues in our tool, also with the help of AIGDD2. The resulting more robust version AMulet2.2 is released with this paper.

Secondly, we aim to deduce the *correctness* of the verification tools using differential testing [24], see Fig. 3. We run both tools AMulet2.2 and DyPoSub on the generated benchmarks that may or may not be correct. If both solvers agree on the correctness of the given circuit, we label the multiplier accordingly and conclude that the tools are correct on these benchmarks. This corresponds to the green boxes in Fig. 3 (top & bottom of the middle column).

However, since we only have two solvers available we automatically have a tie when the solvers disagree on the verification results. Obviously one of the tools has a correctness issue, cf., the red boxes in Fig. 3. We use the following terminology as common in logical reasoning and call a tool *unsound* whenever it decides that a given buggy multiplier is correct. If a correct multiplier is classified as buggy we call the solver *incomplete*.

Since we do not have a golden verification tool available, it is hard to determine which solver is correct and which is buggy without additional actions. The tool DYPOSUB only provides a yes-or-no answer as a verification result, whereas AMULET2 additionally provides certificates. In the case AMULET2 returns "correct circuit" it generates on-the-fly a proof certificate in the practical algebraic calculus (PAC) [15] that independently monitors the reduction steps using a series of simple polynomial operations. The proof certificates can be checked using the proof checkers PACHECK 2.0 or PASTÈQUE 2.0 [15], where the latter is itself certified as it is derived using Isabelle/HOL.

If AMULET2 considers the given circuit to be incorrect, we are able to provide at least one concrete counterexample from the final non-zero reduction result, i.e., we provide a concrete assignment of inputs for which the multiplier produces a wrong result. Due to the reverse topological variable ordering the final reduction polynomial consists only of input variables of the multiplier. We choose the smallest monomial of the polynomial and set all occurring variables to 1, and all other input variables to 0. Hence, the polynomial evaluates to the non-zero coefficient of the selected monomial. In the case we have multiple monomials of minimial size available, we are able to provide several counterexamples. These counterexamples can be used to test the correctness of the multiplier by simulating the outputs, e.g., using AIGSIM [2]. AIGSIM computes the output of the AIG for the provided input model.

So in these cases where AMULET2 and DYPOSUB disagree on the correctness of a circuit, proofs and testing of counterexamples allow us to have a high level of confidence in the verification results of AMULET2. If both tools provide different yes-or-no answers we use the additional tests and proofs by AMULET2. If these tests and proofs are successfully checked by independent simulators and proof checkers, we use the verification result of AMULET2. If the proof certificates and counterexamples are not valid we use the verification result of DYPOSUB and classify AMULET2 as unsound or incompolete.

This tool debugging flow also shows the importance of tests and proofs and how developers of tools that do not provide certificates can make use of counterexamples and proofs of a second tool to debug correctness issues in their tool. Moreover in the case that DYPOSUB would provide counter examples too, providing proofs in AMULET2 becomes superfluous and we could make use of their counter examples to check for unsound errors in AMULET2.

Theoretically we would also need to consider the cases in Fig. 3, where both tools return the same yes-or-no answer, but checking the proofs and counterexamples of AMULET2 fails. However, we have not seen this case in practice, i.e., whenever booth tools returned the same yes-or-no answer, the proofs and counterexamples supported this decision, and thus decided to not include them in Fig. 3.

7   Experiments

Our experiments use an Intel Xeon E5-2620 v4 CPU at 2.10 GHz (with turbo-mode disabled) with a memory limit of 128 GB. The time is listed in seconds (wall-clock time). We compare our tool AMULET2 to related work DYPOSUB, and consider two versions of AMULET2: (i) AMULET2.1 as a slightly updated version of AMULET2.0 [13] that is currently available on GITHUB [10] and (ii) AMULET2.2, a fixed version of AMULET2.1 that will be released together with this paper. For DYPOSUB we use the version described in [21], but, as the tool DYPOSUB is not yet publicly available, we use a binary kindly provided by the authors. In all our experiments the time limit has been set to 300 s and the memory limit has been set to 16 GB.

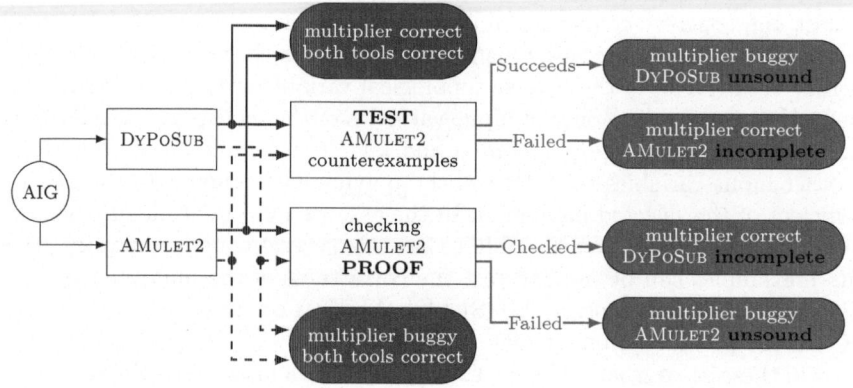

Fig. 3. Determining the correctness of tools using tests and proofs

In our experimental evaluation we aim to investigate the robustness and correctness of the aforementioned multiplier verification tools using benchmarks that are generated by (i) MULTAIGENFUZZER, (ii) AIGOFUZZING on available benchmarks from the AMG-benchmarks [9], and (iii) AIGOFUZZING on benchmarks generated by MULTAIGENFUZZER. Additionally we show the effect of AIGDD2 on reducing the size of failure-inducing inputs. All our experimental data and source code is available at [12].

For each tool we split the results into ✔ and ✗. The ✔-columns show those results where a tool is correct, i.e., where it provides a yes-answer for a correct multiplier (👍), or classifies a faulty multiplier as incorrect (👎). The ✗-columns show failures and miss-classifications, and we use the following symbols to differentiate different types of failures as well as correct results:

✔	✗
👍 Correct multiplier marked as correct	♻ Buggy multiplier marked as correct (unsound)
👎 buggy multiplier marked as incorrect	♺ Cor. multiplier marked as incorrect (incomplete)
	⊀ Static ordering not topological (AMULET2 only)
	⚡ Segmentation fault
	⧖ Exceeding the time limit

7.1 MultAIGenFuzzer Generates Multipliers

The purpose of these experiments is to investigate whether the considered tools have been overfitted to available multiplier benchmarks, e.g. [9, 22, 32]. We use MULTAIGENFUZZER to generate correct multiplier circuits, where the internal structure of the multipliers is shuffled.

In our first experiment that is shown in Table 1, we generate 2 000 random benchmarks with an input bit-width $n \in \{4, 8, 16, 32, 64\}$. In this experiment we exclude the usage of carry-lookahead modules within the circuits, i.e., all benchmarks exist only of full and half adders that are arranged in a random, but topologically correct order. On a first glance this seems to be a rather easy fuzzing setup, however a large number of such benchmarks has been submitted to the SAT Race 2019 [16] and the results show that shuffling the addition order of partial products is hard for SAT solving. We now aim to investigate these benchmarks on algebraic solving techniques.

It can be seen that all tools are able to solve all benchmarks within the given time limit and correctly return that the given benchmarks are indeed correct multipliers. None of the tools has crashes nor returns unsound or incomplete results. Hence, we conclude that these algebraic verification tools are robust with respect to shuffling the addition order of the partial products.

Table 1. MULTAIGENFUZZER benchmarks without carry-lookahead modules.

n	#	AMULET2.1 [10,13] ✔		✗					AMULET2.2 ✔		✗					DYPOSUB [21] ✔		✗			
		👍	👎	♻	♺	⊀	⚡	⧖	👍	👎	♻	♺	⊀	⚡	⧖	👍	👎	♻	♺	⚡	⧖
4	401	401	0	0	0	0	0	0	401	0	0	0	0	0	0	401	0	0	0	0	0
8	414	414	0	0	0	0	0	0	414	0	0	0	0	0	0	414	0	0	0	0	0
16	385	385	0	0	0	0	0	0	385	0	0	0	0	0	0	385	0	0	0	0	0
32	406	406	0	0	0	0	0	0	406	0	0	0	0	0	0	406	0	0	0	0	0
64	394	394	0	0	0	0	0	0	394	0	0	0	0	0	0	394	0	0	0	0	0
	2 000	2 000	0	0	0	0	0	0	2 000	0	0	0	0	0	0	2 000	0	0	0	0	0

In our second experiment we use MULTAIGENFUZZER to generate correct multipliers. For these benchmarks we enable the usage of carry-lookahead modules in the FSA and we want to investigate whether the solvers have been overfitted to certain FSAs. We generate 2 000 random benchmarks with an input bit-width $n \in \{4, 6, 8, 12, 16, 32, 48, 64\}$. We include more input bit-widths than in the first experiment to allow a more detailed discussion. The results can be seen in Table 2, which is organized in the same fashion as Table 1.

The results show that DyPoSub is able to correctly verify all benchmarks within the given time limit. AMulet2.1 as well as the new release AMulet2.2 exceed the time limit on more than 50% of the benchmarks. The number of time-outs increases as the input bit-width increases and AMulet2 is not able to verify any of the 64-bit benchmarks. The reason is that AMulet2 aims to apply adder substitution using syntactic pattern matching before algebraic verification of the rewritten circuit. Since the adder substitution algorithm is not targeted towards mixed FSAs, adder substitution fails in these cases, i.e., algebraic reduction is applied on the original circuit. Hence we are able to conclude that the adder substitution step in AMulet2 is overfitted to known patterns and struggles with random complex inputs.

7.2   Using AIGoFuzzing to Mutate AIGs

In the following experiments we use small 4-bit multipliers that are generated by AMG [9] and use AIGoFuzzing to apply single mutations.

Table 2. MultAIGenFuzzer benchmarks with carry-lookahead modules.

n	#	AMulet2.1 [10,13]							AMulet2.2							DyPoSub [21]					
		✔	✗						✔	✗						✔	✗				
		👍	👎	🖒	🖓	✗	⚡	⏳	👍	👎	🖒	🖓	✗	⚡	⏳	👍	👎	🖒	🖓	⚡	⏳
4	247	247	0	0	0	0	0	0	247	0	0	0	0	0	0	247	0	0	0	0	0
6	221	218	0	0	0	0	0	3	217	0	0	0	0	0	4	221	0	0	0	0	0
8	251	222	0	0	0	0	0	29	225	0	0	0	0	0	26	251	0	0	0	0	0
12	252	141	0	0	0	0	0	111	141	0	0	0	0	0	111	252	0	0	0	0	0
16	249	65	0	0	0	0	0	184	65	0	0	0	0	0	184	249	0	0	0	0	0
32	240	10	0	0	0	0	0	230	10	0	0	0	0	0	230	240	0	0	0	0	0
48	257	1	0	0	0	0	0	256	1	0	0	0	0	0	256	257	0	0	0	0	0
64	283	0	0	0	0	0	0	283	0	0	0	0	0	0	283	283	0	0	0	0	0
	2 000	904	0	0	0	0	0	1 096	906	0	0	0	0	0	1 094	2 000	0	0	0	0	0

First, we select the multiplier circuit "sp-ar-rc-4.aig", that uses a simple partial product generation, i.e., and-gates, which are accumulated in an array structure. The final stage adder is a ripple-carry adder. This architecture consists purely of full and half adders arranged in a grid-like pattern and thus is considered to be a simple multiplier in related work. This circuit is given to AIGoFuzzing where randomly one of four presented mutation techniques is applied.

The results can be seen in Table 3, where the first column refers to the mutation. AIGOFUZZING is able to apply four different kinds of mutations: changing the sign of a random internal edge or node ("intsign"), swapping a random output signal ("outsign"), replacing the input of an internal node ("inptrpl") and integrating a constant ("const+").

We see that changing the internal sign leads to problems with finding an appropriate variable ordering in AMULET2.1. We apply delta debugging on these benchmarks, cf., Table 7 and are able to fix all issues in AMULET2.2. Swapping the sign of an output bit did not produce any crashes nor failures on any of the tools. Modifying the input of an internal node or integrating a constant leads to problems with finding a correct topological variable ordering in AMULET2.1 as well as related segmentation faults. More specifically, integrating a constant leads to crashes of AMULET2.1 on around 50% of the benchmarks.

The tool DYPOSUB is robust and does not crash nor does it exceed the time limit on any of the benchmarks. However, in the row "inptrpl" DYPOSUB reports that 20 circuits are correct, whereas AMULET2.2 only reports 14 circuits to be correct. That is, on 14 circuits both tools return "correct" and on 6 benchmarks DYPOSUB classifies the multiplier to be correct whereas AMULET2 considers the circuit to be buggy. For these 6 circuits we require additional information to resort the tie. We use the counterexamples that are generated by AMULET2.2 and simulate them on the input AIGs. All counterexamples are found to be valid. To be 100% sure of the claim that DYPOSUB is unsound, we additionally generated miters for checking the equivalence of the AIGs, which we gave to a SAT solver. The SAT solver returned SAT, thus the AIGs in question are indeed buggy. Details on the simulations, miters, and results of the SAT solver are included in our experimental data [12]. We select one of the multipliers that causes the unsound-failure. With the help of our delta debugger AIGDD2 we are able to shrink the failure-inducing AIG by 86% from 128 nodes to 18 nodes. We have provided our findings to the authors of DYPOSUB and they were able to locate errors in their tool.

Interestingly, not all mutations where we swapped internal signs and modified the input of nodes lead to incorrect multipliers, which we would have expected. Around 5% of the multipliers were we swap signs, and 2% of the multipliers where we modify inputs of nodes remain correct.

However, we see that only the fourth mutation technique "const+" keeps roughly a 50:50 balance between correct and incorrect multipliers, all other mutations are highly unbalanced. Developing semantics-preserving mutation techniques that preserve correctness of AIGs is part of future work.

We repeated these experimetns of Table 3 with a more complex multiplier "bp-ba-lf-4.aig". This circuit uses a Booth-encoding to generate the partial products, which are then accumulated using a redundant binary addition tree and the FSA is a complex Ladner-Fischer adder. This circuit is modified by AIGO-FUZZING.

These results are shown in Table 4 and it turns out that AMULET2.1 crashes on around 15% of the benchmarks and does not find a correct variable order-

Table 3. AIGoFuzzing "sp-ar-rc-4" multipliers

mutation	#	AMULET2.1 [10,13] ✔ 👍	✔ 👎	✗ 👍	✗ 👎	✗ ⚡̸	✗ ⚡	✗ ⧖	AMULET2.2 ✔ 👍	✔ 👎	✗ 👍	✗ 👎	✗ ⚡̸	✗ ⚡	✗ ⧖	DYPOSUB [21] ✔ 👍	✔ 👎	✗ 👍	✗ 👎	✗ ⚡	✗ ⧖
Intsign	505	31	442	0	0	32	0	0	31	474	0	0	0	0	0	31	474	0	0	0	0
Outsign	524	0	524	0	0	0	0	0	0	524	0	0	0	0	0	0	524	0	0	0	0
Inptrpl	475	14	344	0	0	59	58	0	14	461	0	0	0	0	0	14	455	6	0	0	0
Const+	496	232	0	0	0	12	252	0	252	244	0	0	0	0	0	252	244	0	0	0	0
	2000	277	1309	0	0	103	311	0	297	1703	0	0	0	0	0	297	1702	**6**	0	0	0

ing in around 8% of the cases. However, all of these issues have been resolved in AMULET2.2. Furthermore, DYPOSUB claims that almost all multipliers are incorrect, verifies one circuit and crashes on a second one. On the other hand, AMULET2.2 reports "correct multiplier" on 260 multipliers.

Accordingly, we check the provided proof certificates by AMULET2.2 as support of the verification result. For all multipliers that have been classified as correct by AMULET2.2, the proof certificates are correctly checked by the certified proof checker PASTÈQUE 2.0 [15]. Hence, DYPOSUB is incomplete on these 258 benchmarks.

7.3 Combining Generation- and Mutation-Based Fuzzing

In these experiments we combine our presented fuzzing tools and generate circuits using MULTAIGENFUZZER that are modified using AIGoFuzzing.

Table 4. AIGoFuzzing "bp-ba-lf-4" multipliers

Mutation	#	AMULET2.1 [10,13] ✔ 👍	✔ 👎	✗ 👍	✗ 👎	✗ ⚡̸	✗ ⚡	✗ ⧖	AMULET2.2 ✔ 👍	✔ 👎	✗ 👍	✗ 👎	✗ ⚡̸	✗ ⚡	✗ ⧖	DYPOSUB [21] ✔ 👍	✔ 👎	✗ 👍	✗ 👎	✗ ⚡	✗ ⧖
Intsign	529	23	453	0	0	53	0	0	23	506	0	0	0	0	0	0	506	0	23	0	0
Outsign	476	0	476	0	0	0	0	0	0	476	0	0	0	0	0	0	476	0	0	0	0
Inprpl	544	9	383	0	0	71	81	0	10	534	0	0	0	0	0	0	534	0	9	1	0
Const+	451	193	0	0	0	32	226	0	227	224	0	0	0	0	0	1	224	0	226	0	0
	2000	225	1312	0	0	156	307	0	260	1740	0	0	0	0	0	1	1740	0	**258**	1	0

In the experiment that can be seen in Table 5 we generate benchmarks with an input size $n \in \{4, 6, 8, 12, 16\}$ and apply a single random mutation, i.e., any of the four available types.

We see that AMULET2.1 crashes on $\sim 18\%$ of these benchmarks and has an error in finding a topological variable order in 76 cases. These issues are fixed in AMULET2.2. DYPOSUB is robust and correct on these benchmarks. For all tools we see that the larger the input size the more likely is it to exceed the time limit, i.e., we produce hard benchmarks for these solvers.

Table 5. MULTAIGENFUZZER multipliers + single mutations of AIGOFUZZING

n	#	AMULET2.1 [10,13]							AMULET2.2							DYPOSUB [21]						
		✔	✗						✔	✗						✔	✗					
		👍	👎	👍	👎	✗	⚡	⌛	👍	👎	👍	👎	✗	⚡	⌛	👍	👎	👍	👎	⚡	⌛
4	403	61	253	0	0	19	70	0	66	337	0	0	0	0	0	66	337	0	0	0	0
6	398	41	235	0	0	15	61	46	43	334	0	0	0	0	21	45	352	0	0	0	1
8	426	47	160	0	0	22	93	104	49	256	0	0	0	0	121	55	295	0	0	0	76
12	386	32	112	0	0	6	75	161	31	169	0	0	0	0	185	44	204	0	0	0	138
16	387	38	98	0	0	14	74	163	40	160	0	0	0	0	187	54	184	0	0	0	149
	2 000	219	858	0	0	76	373	474	229	1 257	0	0	0	0	514	264	1 372	0	0	0	364

In the experiment in Table 6 we apply a number of random mutations on a 4-bit multiplier. After the experiment in Table 6 with $n = 4$ we tried various other input sizes and it turns out that already bit-width 6 leads to time outs for a large number of benchmarks. It also shows that increasing the number of mutations also increases the number of crashes in AMULET2.1. Almost 75% of the multipliers either lead to a segmentation fault or an error in the variable ordering. Again, we were able to fix these issues in AMULET2.2. Finally, note, that AMULET2.2 and DYPOSUB do not exceed the time limit on any of these benchmarks and consistently produce the same verification result.

7.4 Using AIGdd2 Minimize Failure-Inducing Inputs

Tables 3, 4, 5 and 6 show that AMULET2.1 is very sensitive with respect to mutations. It produced many crashes and errors while finding an appropriate variable ordering. In order to reduce time consuming debugging during fixing these issues it is beneficial to automatically shrink failure-inducing inputs through delta-debugging. In order to determine the effectiveness of delta-debugging in this context we applied our delta-debugger AIGDD2 with and without slicing on all benchmarks that cause a failure in the variable ordering, i.e., all benchmarks in columns " ✗" of AMULET2.1 in Tables 3, 4, 5 and 6. The results are shown in Table 7.

The table is split into four segments, one for each original table. The rows in each segment are organized in the same fashion as in the original tables, e.g., the first segment considers Table 3, where each row indicates the used mutation type. We also list the number of benchmarks that exceed the time limit of 300 s provided for delta debugging (⌛), benchmarks where delta debugging fails (✗), and where AIGDD2 is able to shrink the input (✔).

It can be seen that AIGDD2 is able to find smaller failure-inducing benchmarks for the multipliers where either internal signs have been swapped or the input of a node has been modified. The columns in block "✔" show the number of benchmarks ("#"), the minimum percentage of nodes that is removed ("min"), the largest percentage that is removed ("max") and the average size removal ("avg"). We see that for the failure-inducing benchmarks of Table 3,

Table 6. MULTAIGENFUZZER multipliers +mutations of AIGOFUZZING

#mut	#	AMULET2.1 [10,13]							AMULET2.2							DyPoSub [21]					
		✔		✗					✔		✗					✔		✗			
		👍	👎	🖒	🖓	⚡̸	⚡	⏳	👍	👎	🖒	🖓	⚡̸	⚡	⏳	👍	👎	🖒	🖓	⚡	⏳
2	191	5	123	0	0	12	51	0	5	186	0	0	0	0	0	5	186	0	0	0	0
4	233	0	120	0	0	18	95	0	0	233	0	0	0	0	0	0	233	0	0	0	0
6	207	0	75	0	0	15	117	0	0	207	0	0	0	0	0	0	207	0	0	0	0
8	223	0	59	0	0	9	155	0	0	223	0	0	0	0	0	0	223	0	0	0	0
10	180	0	42	0	0	3	135	0	0	180	0	0	0	0	0	0	180	0	0	0	0
12	194	0	30	0	0	4	160	0	0	194	0	0	0	0	0	0	194	0	0	0	0
14	181	0	19	0	0	1	161	0	0	181	0	0	0	0	0	0	181	0	0	0	0
16	203	0	18	0	0	4	181	0	0	203	0	0	0	0	0	0	203	0	0	0	0
18	190	0	19	0	0	1	170	0	0	190	0	0	0	0	0	0	190	0	0	0	0
20	198	0	11	0	0	2	185	0	0	198	0	0	0	0	0	0	197	0	0	1	0
	2 000	5	516	0	0	69	1 410	0	5	1 995	0	0	0	0	0	5	1 994	0	0	1	0

AIGDD2 is able to reduce the size by half on average. For the failure-inducing benchmarks of Table 4, AIGDD2 is able to reduce the size by two third.

For those benchmarks, where we integrate constants, AIGDD2 is not able to find smaller inputs that are failure-inducing, cf., column ✗. The reason is in the functionality of AIGDD2. AIGDD2 sets large parts of the AIG to constants and propagates these values, i.e., the internal constant is propagated to the outputs of the AIG. Benchmarks of this kind do not produce a failure in AMULET2.1. A closer inspection of AMULET2.1 showed that we considered multipliers with constant outputs and inputs, but did not consider constants as inputs of internal nodes. Overall we see that slicing-based delta debugging finds smaller failure-inducing benchmarks.

For the benchmarks of Table 5 we see that AIGDD2 several times exceeds the time limit of 300 s for multipliers with an input bit-width $n > 4$. We have repeated the experiment with a time limit of 3 600 s, which shows the same outcome. For those multipliers which could be minimized, AIGDD2 is able to find failure-inducing multipliers that are 96.7% smaller.

For the benchmarks of Table 6 it can be seen that AIGDD2 does not find smaller benchmarks on 10 out of 69 test cases. In all other cases the percentage of removal is between 44.5%–84.2%.

Table 7. Reducing the size of failure-inducing benchmarks with AIGDD2

		⧖	✗	✔				⧖	✗	✔		
	mutation			#	min(%)	max(%)	avg(%)			#	min(%)	max(%)	avg(%)
Table 3	intsign	0	0	32	49.2	53.1	51.7	0	0	32	49.2	68.0	56.5
	inptrpl	0	0	59	48.4	58.3	52.5	0	0	59	48.4	82.0	58.4
	const+	0	12	0	-	-	-	0	12	0	-	-	-
	mutation	⧖	✗	✔				⧖	✗	✔			
				#	min(%)	max(%)	avg(%)			#	min(%)	max(%)	avg(%)
Table 4	intsign	0	0	53	60.2	69.3	65.5	0	0	53	60.2	86.5	67.4
	inptrpl	0	0	71	58.8	73.4	66.1	0	0	71	58.8	88.5	68.5
	const+	0	32	0	-	-	-	0	32	0	-	-	-
	n	⧖	✗	✔				⧖	✗	✔			
				#	min(%)	max(%)	avg(%)			#	min(%)	max(%)	avg(%)
Table 5	4	0	4	15	43.5	63.4	53.0	0	4	15	43.5	63.4	53.0
	6	12	3	0	-	-	-	7	3	5	74.6	89.7	81.8
	8	20	3	0	-	-	-	13	2	7	87.7	94.3	91.0
	12	6	0	0	-	-	-	5	0	1	96.7	96.7	96.7
	16	11	3	0	-	-	-	12	2	0	-	-	-
	#mut	⧖	✗	✔				⧖	✗	✔			
				#	min(%)	max(%)	avg(%)			#	min(%)	max(%)	avg(%)
	2	0	3	9	47.0	59.8	53.8	0	3	9	47.0	70.3	60.4
	4	0	4	14	44.5	71.3	56.5	0	4	14	44.5	71.3	60.0
	6	0	1	14	43.1	67.3	56.1	0	1	14	45.0	72.0	61.4
Table 6	8	0	1	8	47.8	67.2	58.0	0	1	8	54.1	68.8	64.0
	10	0	0	3	55.8	65.0	60.1	0	0	3	55.8	74.8	63.4
	12	0	0	4	54.2	68.5	61.2	0	0	4	58.9	84.2	73.7
	14	0	0	1	56.6	56.6	56.6	0	0	1	56.6	56.6	56.6
	16	0	0	4	52.4	64.2	55.9	0	0	4	52.4	70.6	60.0
	18	0	1	0	-	-	-	0	1	0	-	-	-
	20	0	0	2	56.7	56.8	56.8	0	0	2	56.7	67.8	62.3

(The two major column groups are headed **-slicing** and **+slicing**.)

7.5 Summary

Our experiments have shown that AMULET2 is overfitted to existing FSAs. Especially those MULTAIGENFUZZER-benchmarks, where the FSA is not a clean ripple-carry adder or a pure carry-lookahead adder, but a random combination of both adders are very hard for AMULET2. The tool DYPOSUB is robust and correct on these benchmarks.

We encountered that AMULET2.1 has several issues regarding its robustness on random mutated input circuits. Especially the variable ordering algorithm has severe issues. Delta debugging these failure-inducing inputs was extremely helpful to find the errors in AMULET2.1, which are fixed in the more robust version AMULET2.2.

Using mutation-based fuzzing we have found multipliers where DyPoSub is unsound and others where it is incomplete, i.e., produces incorrect verification results. Since AMulet2 produces checkable proof certificates and counterexamples in addition to a simple yes-or-no answer, it adds another layer of confidence in the verification result and we did not encounter any soundness and completeness failures in AMulet2.

8 Conclusion

Software is only as good as its robustness and correctness. A software that frequently crashes is as undesirable as software that returns incorrect results. In this paper we have evaluated the robustness and correctness of automated multiplier verification tools that use algebraic reasoning. Our presented generation- and mutation-based fuzzing techniques together with the presented delta debugger allow us to detect and debug issues in these tools. Our experiments show that both of the considered tools have critical defects, i.e., DyPoSub [21] is unsound and incomplete, and our tool AMulet2 [13] crashes several times. We fixed the issues in the new release AMulet2.2.

We conclude with observations which we believe to generalize for testing and debugging other verification tools. First, randomly shuffling the structure of available inputs helps to avoid overfitting to known benchmark families. Second, even only small mutations are able to reveal defects in an efficient way. Third, verification tools need to produce proof certificates and models in addition to a yes-or-no answer to prevent false results. Fourth, shrinking failure-inducing inputs using delta debugging is extremely helpful for zooming in on defects.

In the future we want to include dedicated sophisticated semantic preserving mutations in AIGoFuzzing in order to achieve a more balanced ratio between correct and incorrect benchmarks.

References

1. Artho, C., Biere, A., Seidl, M.: Model-based testing for verification back-ends. In: Veanes, M., Viganò, L. (eds.) TAP 2013. LNCS, vol. 7942, pp. 39–55. Springer, Heidelberg (2013). https://doi.org/10.1007/978-3-642-38916-0_3
2. Biere, A., Heljanko, K., Wieringa, S.: AIGER 1.9 And Beyond. Technical report, FMV Reports Series, JKU Linz, Austria (2011)
3. Blotsky, D., Mora, F., Berzish, M., Zheng, Y., Kabir, I., Ganesh, V.: StringFuzz: a fuzzer for string solvers. In: Chockler, H., Weissenbacher, G. (eds.) CAV 2018. LNCS, vol. 10982, pp. 45–51. Springer, Cham (2018). https://doi.org/10.1007/978-3-319-96142-2_6
4. Brummayer, R., Biere, A.: Fuzzing and delta-debugging SMT solvers. In: SMT Workshop, SMT 2009, pp. 1–5, New York, Association for Computing Machinery (2009)
5. Brummayer, R., Lonsing, F., Biere, A.: Automated testing and debugging of SAT and QBF solvers. In: Strichman, O., Szeider, S. (eds.) SAT 2010. LNCS, vol. 6175, pp. 44–57. Springer, Heidelberg (2010). https://doi.org/10.1007/978-3-642-14186-7_6

6. Ciesielski, M.J., Su, T., Yasin, A., Yu, C.: Understanding Algebraic rewriting for arithmetic circuit verification: a bit-flow model. IEEE TCAD **39**(6), 1346–1357 (2019)

7. Godefroid, P., Levin, M.Y., Molnar, D.A.: Automated whitebox fuzz testing. The Internet Society In NDSS (2008)

8. Herfert, S., Patra, J., Pradel, M.: Automatically reducing tree-structured test inputs. In: ASE, pp. 861–871. IEEE Computer Society (2017)

9. Homma, N., Watanabe, Y., Aoki, T., Higuchi, T.: Formal Design of arithmetic circuits based on arithmetic description language. IEICE Trans. **89-A**(12), 3500–3509 (2006)

10. Kaufmann, D.: Amulet 2.1. https://github.com/d-kfmnn/amulet. SHA 8e1838fa4c6d80091869407c44519e2771694b21

11. Kaufmann, D.: Formal Verification of Multiplier Circuits using Computer Algebra. PhD thesis, Computer Science, Johannes Kepler University Linz (2020)

12. Kaufmann, D.: Artifact for fuzzing and delta-debugging and-inverter graph verification tools (2022). http://fmv.jku.at/aigfuzzing_artifact

13. Kaufmann, D., Biere, A.: AMULET 2.0 for verifying multiplier circuits. In: TACAS 2021. LNCS, vol. 12652, pp. 357–364. Springer, Cham (2021). https://doi.org/10.1007/978-3-030-72013-1_19

14. Kaufmann, D., Biere, A., Kauers, M.: Verifying SAT and computer algebra. In: FMCAD 2019, pp. 28–36. IEEE (2019)

15. Kaufmann, D., Fleury, M., Biere, A., Kauers, M.: Practical algebraic calculus and nullstellensatz with the checkers pacheck and pastèque and nuss-checker. Formal Methods Syst. Des., 35 p. (2022, online first). https://doi.org/10.1007/s10703-022-00391-x

16. Kaufmann, D., Kauers, M., Biere, A., Cok, D.: Arithmetic verification problems submitted to the SAT race 2019. In: SAT Race 2019, volume B-2019-1 of Dep. of Computer Science Report Series B, p. 49. University of Helsinki, (2019)

17. Kremer, G., Niemetz, A., Preiner, M.: ddSMT 2.0: better delta debugging for the SMT-LIBv2 language and friends. In: Silva, A., Leino, K.R.M. (eds.) CAV 2021. LNCS, vol. 12760, pp. 231–242. Springer, Cham (2021). https://doi.org/10.1007/978-3-030-81688-9_11

18. Kuehlmann, A., Paruthi, V., Krohm, F., Ganai, M.: Robust Boolean reasoning for equivalence checking and functional property verification. IEEE TCAD **21**(12), 1377–1394 (2002)

19. Lampropoulos, L., Hicks, M., Pierce, B.C.: Coverage guided, property based testing. Proc. ACM Program. Lang. **3**(OOPSLA), 181:1–181:29 (2019)

20. Mahzoon, A., Große, D., Drechsler, R.: RevSCA: using reverse engineering to bring light into backward rewriting for big and dirty multipliers. In: DAC 2019, pp. 185:1–185:6. ACM (2019)

21. Mahzoon, A., Große, D., Scholl, C., Drechsler, R.: Towards formal verification of optimized and industrial multipliers. In: DATE, pp. 544–549. IEEE (2020)

22. A. Mahzoon, D. Große, and R. Drechsler. Multiplier Generator GenMul (2019). http://www.sca-verification.org/

23. Mansur, M.N., Christakis, M., Wüstholz, V., Zhang, F.: Detecting critical bugs in SMT solvers using blackbox mutational fuzzing. In: ESEC/SIGSOFT FSE, pp. 701–712. ACM (2020)

24. McKeeman, W.M.: Differential testing for software. Digit. Tech. J. **10**(1), 100–107 (1998)

25. B. Miller, M. Zhang, and E. Heymann. The relevance of classic fuzz testing: have we solved this one? IEEE Trans. Softw. Eng. 1 (2020, early acces). https://doi.org/10.1109/TSE.2020.3047766
26. Miller, B.P., Fredriksen, L., So, B.: An empirical study of the reliability of unix utilities. Commun. ACM **33**(12), 32–44 (1990)
27. Niemetz, A., Biere, A.: ddSMT: a delta debugger for the SMT-LIB v2 Format. In: Bruttomesso, R., Griggio, A. (eds.), Proceedings of the 11th International Workshop on Satisfiability Modulo Theories, SMT 2013), affiliated with the 16th International Conference on Theory and Applications of Satisfiability Testing, SAT 2013, Helsinki, Finland, 8–9 July 2013, pp. 36–45 (2013)
28. Niemetz, A., Preiner, M., Biere, A.: Model-based API testing for SMT solvers. In: SMT, volume 1889 of CEUR Workshop Proceedings, pp. 3–14. CEUR-WS.org (2017)
29. Parhami, B.: Computer Arithmetic - Algorithms and Hardware designs. Oxford University Press, Oxford (2000)
30. Scott, J., Sudula, T., Rehman, H., Mora, F., Ganesh, V.: BanditFuzz: fuzzing SMT solvers with multi-agent reinforcement learning. In: Huisman, M., Păsăreanu, C., Zhan, N. (eds.) FM 2021. LNCS, vol. 13047, pp. 103–121. Springer, Cham (2021). https://doi.org/10.1007/978-3-030-90870-6_6
31. Sharangpani, H., Barton, M.L.: Statistical analysis of floating point flaw in the pentium processor. In: Technical Report Intel Corporation (1994)
32. Temel, M.: MultGen. https://github.com/temelmertcan/multgen (2020). SHA 32f4eb1bff419a88f02035d365d88089f6c33e5f
33. Zeller, A.: Yesterday, my program worked. today, it does not. why? In: Nierstrasz, O., Lemoine, M. (eds.) ESEC/SIGSOFT FSE -1999. LNCS, vol. 1687, pp. 253–267. Springer, Heidelberg (1999). https://doi.org/10.1007/3-540-48166-4_16
34. Zeller, A.: The Debugging Book. CISPA Helmholtz Center for Information Security (2021)
35. Zeller, A., Gopinath, R., Böhme, M., Fraser, G., Holler, C.: The Fuzzing Book. CISPA Helmholtz Center for Information Security (2021)
36. Zeller, A., Hildebrandt, R.: Simplifying and isolating failure-inducing input. IEEE Trans. Softw. Eng. **28**(2), 183–200 (2002)

A Unit-Based Symbolic Execution Method for Detecting Heap Overflow Vulnerability in Executable Codes

Maryam Mouzarani$^{(\boxtimes)}$ ⓘ, Ali Kamali ⓘ, Sara Baradaran ⓘ,
and Mahdi Heidari ⓘ

Department of Electrical and Computer Engineering,
Isfahan University of Technology, Isfahan, Iran
mouzarani@iut.ac.ir, {a.kamali,s.baradaran,heidari}@ec.iut.ac.ir

Abstract. Symbolic execution has been a popular method for detecting vulnerabilities of programs in recent years, yet path explosion has remained a significant challenge in its application. This paper proposes a method for improving the efficiency of symbolic execution and detecting heap overflow vulnerability in executable codes. Instead of applying symbolic execution to the whole program, our method initially determines test units of the program, which are parts of the code that might contain heap overflow vulnerability. This is performed through static analysis and based on the specification of heap overflow vulnerability. Then, it applies symbolic execution to the test units and extracts a constraint tree for each unit. Every node in this tree contains the path and vulnerability constraints on the unit input data for executing and overflowing heap buffers in that node. Solving these constraints gives us input values for the test unit that execute the desired nodes and cause heap overflow. Finally, we use curve fitting and treatment learning to approximate the relation between system and unit input data as a function. Using this function, we generate system inputs that enter the program, reach vulnerable instructions in the desired test unit, and cause heap overflow in those instructions. This method is implemented as a plugin for *angr* framework and evaluated using a group of benchmark programs. The experiments show its superiority over similar tools in accuracy and performance.

Keywords: Unit testing · Symbolic execution · Executable codes · Heap overflow · Machine learning

1 Introduction

A wide variety of program analysis and vulnerability detection techniques have been introduced in the past decades, among which symbolic execution has attracted a great deal of attention [10]. Although symbolic execution is theoretically sound and complete [4], it may run into challenges in analyzing real-world programs, such as path explosion. Here, the number of program execution

© The Author(s), under exclusive license to Springer Nature Switzerland AG 2022
L. Kovács and K. Meinke (Eds.): TAP 2022, LNCS 13361, pp. 89–105, 2022.
https://doi.org/10.1007/978-3-031-09827-7_6

paths may grow exponentially, making storing and exploring the program paths impractical. Some solutions have been proposed to overcome this challenge such as pruning infeasible paths [20], function and loop summarization [13,14], state merging [13,15], and compiler optimizations [12].

Some researchers have applied machine learning methods to improve symbolic execution and contain path explosion [6–9,11,19]. For instance, in [11], symbolic execution is simply applied to a given program unit rather than the entire program to limit the scope of symbolic analysis and avoid path explosion. In this method, symbolic execution is used to analyze the constraints of execution paths in the unit and calculate appropriate unit input data covering all paths in the test unit. Then, the curve fitting technique [3] is employed to approximate the relationship between system inputs and the given test unit input data. Finally, system inputs are generated that are correlated to the calculated unit input data. This method is not used to detect a specific class of vulnerability, and it does not contain details on how to determine the test units in a program.

This paper extends the idea presented in [11] and proposes a method for detecting heap overflow vulnerability in executable codes. Our method clearly defines how to automatically determine test units in executable code according to our specification of heap overflow vulnerability. We apply symbolic execution to each unit and, given the specification of heap overflow vulnerability, calculate path and vulnerability constraints in each execution path of the unit. We generate unit input data to explore a test unit, reach the vulnerable statements, and cause heap overflow by solving the calculated constraints. Similar to the method in [11], we estimate the relationship between the program input data and that of the test unit by simulating the program execution and using machine learning techniques. In this way, we generate test data that enters into the program from the beginning and activates vulnerability in the desired instruction of the test unit.

Our method has been implemented as a plugin for *angr* framework [23] and is available in [1]. We have evaluated the performance and accuracy of our method using NIST SARD benchmark vulnerable programs [2] and a designed complex program, presented in Listing 2, that contains more functions and more complicated path constraints compared to the benchmark programs. Our solution has been compared with two similar heap overflow detection tools named MACKE [18] and Driller [21]. The experiments show that our method performs more efficiently and accurately than these tools for detecting heap overflow vulnerability.

To summarize, our contributions are as follows:

- Specifying heap overflow vulnerability in executable codes and presenting a method to automatically determine test units in a program accordingly
- Revising the testing algorithm presented in [11] to focus on detecting heap overflow vulnerability more efficiently
- Implementing and evaluating the total solution to demonstrate the advantages of unit-based symbolic execution against similar methods for detecting heap overflow vulnerability

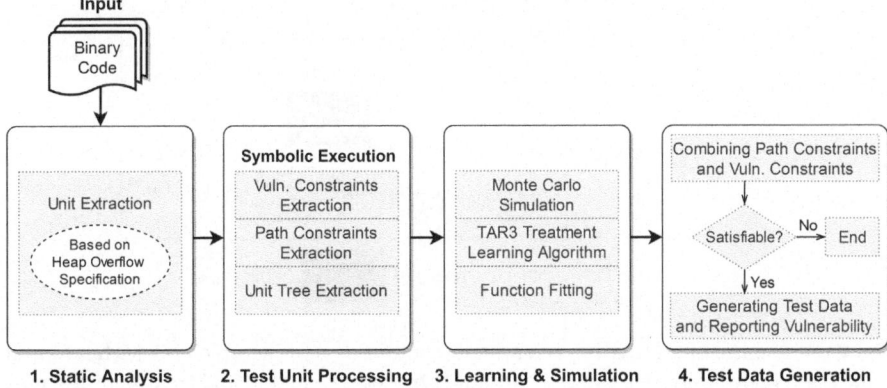

Fig. 1. Architecture of the proposed method

The remainder of this paper is structured as follows: In Sect. 2, the proposed method is described in detail. Section 3 evaluates the implemented method, and finally, Sect. 4 concludes the paper and presents some future works.

2 Method Overview

Our proposed method consists of four major phases, as illustrated in Fig. 1. In the first phase, the program executable code is statically analyzed to identify test units based on the specification of heap overflow vulnerability. To make the process clearer, Fig. 2 illustrates a program containing various possibly vulnerable units for which we explain the steps of our proposed solution briefly. Our method recognizes the test unit, shown in black, statically according to the specification of heap overflow vulnerability. In fact, we are interested in finding unit input data i_k and its relevant system input data I_k that causes heap overflow in the unit. In the second phase, we analyze all execution paths in the unit through symbolic execution and consider the rest of the program as a black box. More precisely, we perform symbolic execution in this phase and create a constraint tree for the extracted unit that contains path and vulnerability constraints on unit input data for each possibly vulnerable statement. In the third phase, Monte Carlo simulation is performed, and the whole program is executed with multiple system input values. If system input I_k reaches the test unit with input value i_k and causes the execution of a node n in the unit tree, we annotate node n with the pair (I_k, i_k) to record which input data causes executing that node. Then, for each possibly vulnerable node in the unit tree, we use function fitting technique [22] to estimate the relation between system and unit input data as a function. Finally, in the fourth phase, we use the calculated path and vulnerability constraints and the estimated function to generate system input data that enters the program, reaches the test unit, and causes heap overflow in vulnerable statements. In the following, we explain each phase in more detail.

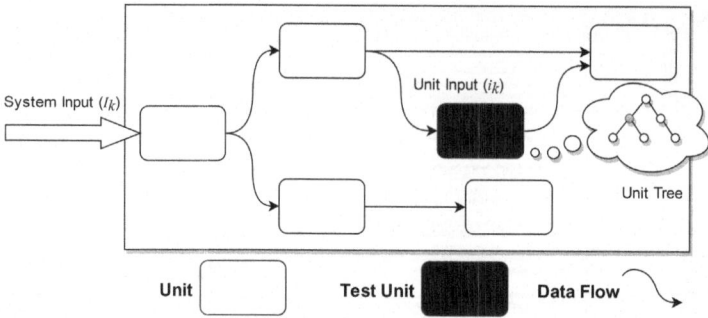

Fig. 2. Schematic of a program as a system containing a vulnerable unit with an input i_k having a relevant system input I_k obtained from curve fitting and treatment learning

2.1 Static Analysis

In the first phase, we analyze the program executable code statically and search for functions that might contain heap overflow vulnerability. To locate possibly vulnerable statements in executable code, we first specify how heap overflow vulnerability appears. We use the general vulnerability specification method presented in [17] to describe vulnerabilities as a sequence of pairs $\{CONT, Rule\}$ that specify the data concerned in a vulnerability and the conditions on it for the vulnerability activation, respectively. We describe heap overflow using this method and based on VEX language since our proposed solution is implemented as a plugin for *angr*, that translates binary instructions into VEX intermediate language. Figure 3 presents our specification of heap overflow as a multi-event vulnerability.

This specification consists of two events: allocating a heap buffer and storing some data into that buffer. Actually, the symbol ▷ represents the sequence of two events. In this specification, the following containers are considered:

- $CONT1$: address of the allocated heap memory (the address returned from the *malloc* function)
- $CONT2$: length of the allocated heap memory (the input argument of the *malloc* function which is assumed to be a constant value)
- $CONT3$: address on which arbitrary data is stored using a *store* instruction
- $CONT4$: data stored using a *store* instruction

According to the *Rule* illustrated in the second part of the specification, this vulnerability occurs when some data is stored in a heap buffer using the *Ist_Store* VEX instruction. The source and destination of the *store* operation are defined in the *Ist_Store.data.tmp* and *Ist_Store.addr* sections of this VEX instruction, respectively. If length of the source data is more than size of the destination heap buffer, or if $length(CONT4)$ is greater than $CONT2$, then heap overflow occurs. According to this specification, we identify allocated heap buffers ($CONT1$) in

$$(\{CONT1, CONT2\}, True) \vartriangleright (\{CONT3, CONT4\}, Rule)$$

1. $CONT1 = malloc(CONT2)$
2. $CONT3 = Ist\_Store.addr$
3. $CONT4 = Ist\_Store.data.tmp$

$Rule:$
$$CONT1 \leq CONT3 < CONT1 + CONT2$$
$$and$$
$$len(CONT4) > CONT2$$

Fig. 3. Our specification of heap overflow vulnerability in VEX language

executable codes and search for functions in which some data is stored in these buffers. Such functions are considered as test units. It is worth mentioning that since the address of a heap buffer is dynamically determined at run time, we use the local variable that stores the address of the allocated heap buffer. Since this variable is located in the stack memory, we use it to follow the usage of that buffer through the program statements and functions. One of our challenges in recognizing test units in executable codes is following the usage of the heap buffer in nested function calls. Under such circumstances, the heap buffer address may be sent to other functions as an argument. To overcome this challenge, we use calling conventions to follow the local variable holding address of the heap buffer, which is passed to other functions or returned from function calls. We also assume that the length of the allocated heap buffer in the *malloc* function is a constant value, and it is available during static analysis.

2.2 Processing the Test Units

In this phase, symbolic execution is applied to each unit after determining the test units using static analysis, and a constraint tree is extracted. In this tree, each node is annotated with metadata obtained from symbolic execution that shows the system state at that point of the program. This metadata contains the path constraints from the beginning of the test unit to the given node, the node constraints, and the vulnerability constraints in that node. Vulnerability constraints are calculated based on the length of the heap buffer for nodes in which some data is stored into a heap buffer. These constraints are according to the vulnerability specification in Fig. 3.

2.3 Learning and Simulation Process

After extracting the constraint tree, the program execution is simulated, and its behavior is learned in the third phase of our solution. Details of the operations in this phase are presented in Algorithm 1. This algorithm is a revised version of

---

Algorithm 1. $Cover(S, U, T)$

Input: System S with inputs I with $d = |I|$, unit U with inputs i and constraint tree T obtained from applying symbolic execution to the unit U

1: Perform n-factor combinatorial MC simulation over space R^d
2: $(V, v) \leftarrow \{(a, b)|$ a is a system level vector and b is the corresponding monitored unit level vector$\}$
3: **for** node n in T using BFS **do**
4:     **if** n is in a possibly vulnerable path **then**
5:         **if** n and n'siblings are covered **then**
6:             $V' \leftarrow \{a \in V|$ a cover $n\}$ and $V'' \leftarrow V \backslash V'$
7:             $(I_n, R_n, \_) \leftarrow RunTar3(I, V, V', V'')$
8:             $\forall j \in I_n$ store the range $r_j \in R_n$ for j
9:         **else if** n is satisfiable but not covered **then**
10:             $i \leftarrow$ model for $Const(n)$
11:             $C \leftarrow Term(n)$
12:             $(I_n, i_n, f_n) \leftarrow ComputeMap(C, V, v, n, Parent(n), i)$
13:         **end if**
14:     **end if**
15: **end for**
16: **for** node n in T that n is possibly vulnerable and satisfiable **do**
17:     **if** n is covered **then**
18:         Generate input using $Const(n), VulConst(n)$ and $\forall j \in I \backslash I_n$ use $r_j \in R_n$
19:     **else**
20:         Generate input using $Const(n), VulConst(n)$ and f_n
21:     **end if**
22: **end for**

---

the *Cover* algorithm presented in [11], and our modifications are shown in blue. In this algorithm, the terms $Term(n)$, $Const(n)$, and $VulConst(n)$ refer to the node constraints, the path constraints from the beginning of the test unit to the given node n, and the vulnerability constraints of node n, respectively.

In this algorithm, first in lines 1 and 2, we perform n-factor Monte Carlo simulation on the program and generate possible combinations of input data. We execute the program with these inputs and monitor it to annotate nodes of the unit tree with the system and unit data pairs (I_k, i_k) that reach the aforementioned nodes during the program execution.

Next, in lines 3 and 4, we explore the constraint tree and analyze only nodes located in a possibly vulnerable path, a path from the root to a leaf in the constraint tree that contains some nodes in which a *store* operation to a heap buffer is performed. In contrast to the algorithm in [11], which processes all nodes in the constraint tree at this step, we limit our analysis to a group of nodes according to the vulnerability specification to improve the efficiency of our method. In lines 5 to 8, we check if these nodes and their siblings have been executed during the simulation, then we use *TAR3* treatment learning algorithm [16] to estimate the range of system inputs that could explore the desired node in the unit. Otherwise, in lines 9 to 12, for each uncovered node whose path constraints are satisfiable, a function named *ComputeMap* is called that estimates the relation between system and unit input data as a function f. The algorithm of this function is presented in Algorithm 2. By solving the path constraints of the node, we generate a unit input data that reaches the node. We give this data to the function f, and it returns appropriate system input that could explore the desired node in the unit tree during the execution.

2.4 Test Data Generation

Until now, we have only considered path constraints in generating system input data. In lines 16 to 22 of Algorithm 1, for each node containing a possibly vulnerable statement whose path constraints are satisfiable, we try to generate system input data consistent with both path and its calculated vulnerability constraints. To do so, in lines 17 and 18, for each node that has been covered in the simulation step, we solve the path and vulnerability constraints of the node and generate appropriate unit input data to cause overflow in that node. In the last step, using the range of system inputs ($r_j \in R_n$) calculated by *TAR3* algorithm, we find relevant system input data for the desired unit input data.

Next, in lines 19 and 20, for each node that has not been covered in the simulation step, we use the fitted function f to find relevant system input data for the unit input data consistent with calculated path and vulnerability constraints.

To summarize the difference between our *Cover* algorithm and the one presented in [11], first in line 4, we improve the performance of our analysis by only considering nodes in potentially vulnerable paths, while in [11], all the nodes are analyzed in this step even though they might not contain any vulnerability. Next, we calculate both path and vulnerability constraints, and this is statically performed using symbolic execution. However, the algorithm in [11] only considers the path constraints calculated gradually using dynamic symbolic execution by generating new input data that explores uncovered paths in the unit. Thus, we calculate the constraints more quickly. Since our symbolic analysis is restricted to a single function, dynamic symbolic execution accuracy and coverage advantages over symbolic execution are not significant here. Finally, we consider the path and vulnerability constraints in lines 16 to 22 for generating system input data that reaches the vulnerable nodes and causes heap overflow. In contrast, the algorithm in [11] only considers the path constraints for generating system input data that covers the nodes of the unit.

ComputeMap Algorithm. We have used the same algorithm, shown in Algorithm 2, as introduced in [11] for the *ComputeMap* function. We describe this algorithm here to make the whole process clear for the reader. Due to the complexity of applying curve fitting to a large set of data and the presence of a large number of parameters, the algorithm fits the program behavior into a function by initially considering the constraints of each node ($Term(n)$) individually. Thus, the unit input variables related to the node constraints and the constraints applied to each variable are first extracted. Based on these variables, a subset

---

Algorithm 2. ComputeMap(C, I, V, v, n, n', i)

Input: Constraint C, System vectors V, Unit vectors v, a node n that we want to cover, a node n' that is in the parent hierarchy of n and a model i for $Const(n)$
Output: (I_n, i_n, f_n) where $i_n = Vars(C)$ and $I_n = f(i_n)$
1: $i_n = Vars(C)$
2: $i_n =$ restriction of i to i_n
3: $V' \leftarrow \{a \in V |\ a$ is in 20% of points closet to $Const(n)\}$ and $V'' \leftarrow V \backslash V'$
4: $(I_n, R_n, Smooth) \leftarrow$ RunTar3(I, V, V', V'')
5: **if** $Smooth$ **then**
6:      Build map $I_n = f(i_n)$ ▷ curve fitting step
7: **else**
8:      **if** n' exists **then**
9:          $C \leftarrow C \wedge Term(n')$
10:         $(I_n, i_n, f_n) \leftarrow$ ComputeMap($C, I, V, v, n, Parent(n'), i$)
11:     **else**
12:         $n'' \leftarrow n$
13:         **while** $Parent(n'')$ exists **do**
14:             $C \leftarrow C \wedge Term(Parent(n''))$
15:             $n'' \leftarrow Parent(n'')$
16:             $V' \leftarrow \{a \in V |\ a$ cover $n''\}$
17:             **if** $|V'| \geq Threshold$ **then**
18:                 break
19:             **end if**
20:         **end while**
21:         $V'' \leftarrow V \backslash V'$
22:         $(I_n, R_n, \_) \leftarrow$ RunTar3(I, V, V', V'')
23:         $i_n = Vars(C)$
24:         Build map $I_n = f(i_n)$ ▷ curve fitting step
25:     **end if**
26: **end if**

---

of unit inputs for which there are path constraints is extracted. Then, the path constraints associated with these inputs are calculated. In the next step, the first 20% of system inputs that are more compatible with this constraints subset are selected as a set V'. Afterward, *TAR3* algorithm is applied to the set, and if a smooth[1] relationship is established there between, the function f is built using curve fitting. Otherwise, the process is repeated recursively by adding parent node constraints to the given node in order to establish a smooth relationship.

If a smooth relationship is not found by including all terms in $Const(n)$, in lines 11 to 24 we walk up through the unit tree to find a parent node with enough system input values in its annotation. Such node is covered in the simulation step with appropriate number of input data (I_k, i_k) that helps to better estimate the function f using the curve fitting algorithm.

---

[1] The smoothness of a function is a stronger case than the continuity of the function. A *smooth function* is a function having continuous derivatives up to a specific order.

3   Evaluation

The proposed method has been implemented as a plugin for *angr*, a symbolic execution framework for binary analysis [23]. In our implementation, *string* data type is also supported for detecting heap overflow in *string manipulation* functions in addition to *int, short, unsigned int, char, float, double*, and *enum* data types supported in the proposed approach in [11].

We have designed two experiments to evaluate our solution; in the first experiment, a set of 90 programs from NIST SARD benchmark [2] containing heap overflow vulnerability has been used. The vulnerability occurs in these programs when some constant data is copied into a heap buffer using *strcpy, strcat, memcpy*, and *memmove* functions. More precisely, out of 90 test programs, 15 programs have the vulnerability in calling *strcpy* function, 15 programs in calling *strcat*, 30 programs in calling *memcpy*, and 30 programs in calling *memmove*. To better evaluate our proposed method, we have made the path constraints in the test programs more complicated by adding an additional *if* statement to the vulnerable paths. In addition, instead of copying constant data into a heap buffer, we have copied an input variable into that buffer to create a vulnerability constraint in the test unit. A vulnerable function in one of these benchmark programs is presented in Listing 1 as an example, and our added *if* statement is underlined in line 16. The same *if* statement has been similarly added to all benchmark programs. In the second experiment, we have created a test program with several functions and more complicated path and vulnerability constraints to better evaluate the efficiency of our method. The source code of this program, along with its details, is presented in Listing 2.

In both experiments, we have compared our implemented solution with two other tools that use the similar method for detecting vulnerabilities in C programs, namely MACKE [18] and Driller [21].

Driller is a fuzzing tool that uses evolutionary algorithms to generate multiple input values from an initial seed and explore the program paths. If the process is trapped in a part of the program because of a conditional statement and the fuzzer fails to generate consistent input values for that condition, symbolic execution is applied to calculate the constraint and generate appropriate data. Then, the fuzzer generates input data based on the obtained data from symbolic execution to detect more in-depth vulnerabilities. Our proposed method is compared with this tool as it applies symbolic execution and uses *angr* framework to improve the coverage of program analysis. Driller is also among the most popular vulnerability detection tools, given its satisfactory performance in detecting vulnerabilities [21].

MACKE is a framework written on top of the KLEE symbolic execution engine [5] for compositional analysis of C programs [18]. In this framework, the program is divided into different units, and symbolic execution is performed for detecting vulnerabilities in each unit. It recognizes each function through static analysis and considers it as a test unit. MACKE also analyzes the call graph and the program control flow graph statically to identify possible function call scenarios. This way determines whether a function containing vulnerable state-

```
1  // Filename: CWE122_Heap_Based_Buffer_Overflow__c_CWE193_char_cpy_34.c
2  void CWE122_Heap_Based_Buffer_Overflow_char_cpy_34_bad(char * source)
3  {
4      char * data; = NULL;
5      struct fp * ptr = NULL;
6      CWE122_Heap_Based_Buffer_Overflow_char_cpy_34_unionType myUnion;
7      // FLAW: Did not allocate space based on the source length
8      data = (char *)malloc(20*sizeof(char));
9      if(data == NULL) { exit(-1); }
10     ptr = (struct fp *)malloc(sizeof(struct fp));
11     if(ptr == NULL) { exit(-1); }
12     myUnion.unionFirst = data;
13     {
14         char * data = myUnion.unionSecond;
15         ptr->fptr = printLine;
16         if(source[0] == '7' && source[1] == '/' && source[2] == '4'
           && source[3] == '2' && source[4] == 'a' && source[5] == '8'
           && source[75] == 'a')
17         {
18             /* POTENTIAL FLAW:
19             data may not have enough space to hold source */
20             strcpy(data, source);
21         }
22         ptr->fptr("That's OK!");
23         printLine(data);
24         free(data);
25         free(ptr);
26     }
27 }
```

Listing 1. A sample vulnerable unit in benchmark programs

ments may be executed in a sequence of possible function calls. After performing symbolic execution in each test unit and calculating appropriate input data that reveals vulnerabilities in the unit, the tool reports possible vulnerabilities in each unit with the relevant unit inputs as the proof of concept. Since MACKE does not consider the constraints of the path from the beginning of the program to the test unit, it may generate several false positives in this step. Therefore, the last analysis step removes alarms related to the units recognized as unreachable during the static analysis. However, since this analysis is only based on the possible function call scenarios, our evaluations demonstrate false positive and negative alarms in MACKE outputs regardless of the path constraints.

3.1   Experiment 1

Table 1 shows the results of our first experiment for testing NIST SARD benchmark programs. The columns in this table represent, from left to right, the number of true positives (TP), true negatives (TN), false positives (FP), false negatives (FN), and the accuracy metric which is calculated as shown in (1).

Table 1. The results of evaluating the approaches on the benchmark programs

Tool	TP	TN	FP	FN	Accuracy
Driller	90	116	0	0	1.00
MACKE	54	108	8	36	0.78
Our method	90	116	0	0	1.00

$$Accuracy = \frac{TP + TN}{TP + TN + FP + FN} \tag{1}$$

Each test program in NIST SARD contains a vulnerable statement in a function whose name contains the word *"bad"* and one or two functions whose names contain the word *"good"* that use similar statements without vulnerabilities. Thus, a precise tool is expected to have one true positive alarm and one or two true negative alarms for each test program. As shown in Table 1, our tool and Driller could precisely detect all vulnerabilities in the test programs with no false alarms. However, MACKE had 8 false positive and 36 false negative alarms in analyzing the benchmark programs.

We have also compared the execution time of the tools in this experiment, as shown in Fig. 4. This figure shows the average analysis time that each tool has spent on the test programs with a vulnerable function, e.g., *strcpy*, *memcpy*, etc. As observed, the performance of our proposed method has significant superiority over the Driller's. Although the analysis time of MACKE in this experiment has been less than that of our tool, it has generated more false alarms and less accurate results. Additionally, MACKE only generates local input data for executing a single unit and does not consider the path constraints on system input data for reaching the beginning of a test unit. On the contrary, our proposed method generates accurate test data for executing the whole program from the beginning and reaching the vulnerable statement in the test unit. This might be one reason that testing and analyzing a program takes more time in our method.

3.2 Experiment 2

In the second experiment, we have analyzed our designed test program with six vulnerable statements, presented in Listing 2, using all three tools. The designed complex program is a simple authentication code by which the user can carry out the sign-up and sign-in operations. The user should run the program and enter the username and password in the console to sign-up. If the condition in line 115 is satisfied, the vulnerable function *signup* would be called. In this function, two heap buffers are allocated in lines 9 and 11. As there are two copy operations with *memcpy* function calls in lines 14 and 17, this function is identified as a test unit by our solution. There is a path constraint, in line 6, in this function, therefore if the input strings for username and password satisfy the path constraints in lines 115 and 6, and the length of them be more than the length of the destination heap buffers in the copy operations, they would

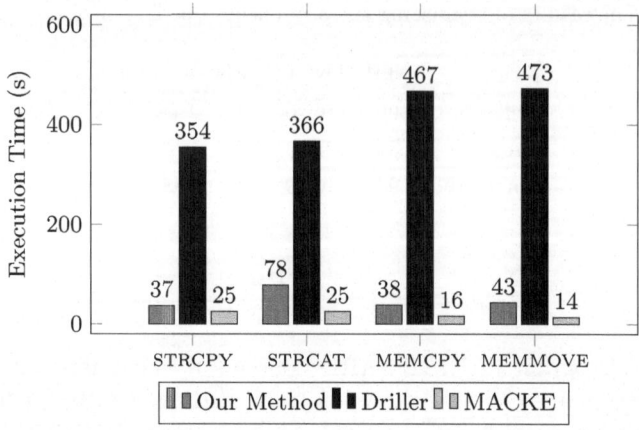

Fig. 4. Comparison of tools analysis time on the benchmark programs

cause heap overflow. Note that the path constraint in line 115 is out of the test unit and should be recognized through machine learning. Our implemented solution calculates the path constraint in line 6 using symbolic execution since this constraint is inside the unit. Then, it generates appropriate input data for the *scanf* operations in lines 113 and 114, consistent with both path constraints inside and outside the unit. There are two other test units in this program, *check* and *authentication* functions, that cause heap overflow by calling *strcpy* and *memcpy* functions respectively. The same challenge exists in these functions for our solution to calculate the path constraints inside the test unit and estimate the ones outside of it.

The results of this experiment are demonstrated in Table 2. As shown in this table, MACKE could detect only one of these vulnerabilities in the given program as it seems that could not analyze complicated path conditions. Driller could detect four vulnerabilities in the test program, and it has generated two false negative alarms. In contrast, our tool has detected all six vulnerabilities precisely. It has generated appropriate test data for the whole program that enters the program from the beginning and causes heap overflow in the vulnerable statements of the test units. Besides, the testing time in our tool has been much less than that of Driller. Therefore, the results of this experiment demonstrate the advantage of restricting the scope of symbolic execution for detecting a specific vulnerability class.

Table 2. The results of evaluating the approaches on the complex test program

Tool	TP	TN	FP	FN	Time(s)
Driller	4	0	0	2	3549
MACKE	1	0	0	5	335
Our method	6	0	0	0	374

```
1 #define def_user "admin"
2 #define def_pass "password"
3 // The function "signup" as a vulnerable unit
4 void signup(char *username, char *password)
5 {
6    if((username[1] >= '0' && username[1] <= '9')
     && !strncmp(password, "passW0rd", 8))
7    {
8       // FLAW: Did not allocate space based on the username length
9       char *tmp_user = (char *)(malloc(50*sizeof(char)));
10      // FLAW: Did not allocate space based on the password length
11      char *tmp_pass = (char *)(malloc(50*sizeof(char)));
12      /* POTENTIAL FLAW:
13      data may not have enough space to hold source */
14      memcpy(tmp_user, username, strlen(username));
15      /* POTENTIAL FLAW:
16      data may not have enough space to hold source */
17      memcpy(tmp_pass, password, strlen(password));
18      if(strlen(tmp_pass) < 12)
19      {
20         printf("The selected password is too weak\n");
21         return;
22      }
23      int fd = open(tmp_user, O_WRONLY|O_CREAT, 0777);
24      if(fd < 0)
25      {
26         printf("An unexpected problem occurred!\n");
27         return;
28      }
29      write(fd,tmp_pass, sizeof(tmp_pass));
30      printf("%s your registration was successful\n", tmp_user);
31   }
32   else if(!(username[1] >= '0' && username[1] <= '9'))
33      printf("The second letter of username must be a number\n");
34   else
35      printf("The password must start with the word <passW0rd>\n");
36 }
37 // The function "check" as a vulnerable unit
38 bool check(char *username, char *password)
39 {
40    // FLAW: Did not allocate space based on the username length
41    char *tmp_user = (char *)(malloc(50*sizeof(char)));
42    // FLAW: Did not allocate space based on the password length
43    char *tmp_pass = (char *)(malloc(50*sizeof(char)));
44    if((username[0] >= 'A' && username[0] <= 'Z')
      && (username[1] >= '0' && username[1] <= '9'))
45    {
46       /* POTENTIAL FLAW:
47       data may not have enough space to hold source */
48       strcpy(tmp_user, username);
49       /* POTENTIAL FLAW:
50       data may not have enough space to hold source */
51       strcpy(tmp_pass, password);
```

```
52        if(!strcmp(tmp_user, def_user) && !strcmp(tmp_pass, def_pass))
53            return true;
54        char passwd[50];
55        int fd = open(tmp_user, O_RDONLY);
56        if(fd < 0)
57        {
58            printf("An unexpected problem occurred!\n");
59            return false;
60        }
61        read(fd, passwd, sizeof(passwd));
62        if(!strcmp(passwd, tmp_pass)) { return true; }
63    }
64    return false;
65 }
66 // The function "signin" without any vulnerable statement
67 bool signin(char *username, char *password)
68 {
69    if(check(username, password))
70    {
71        printf("%s you logged in successfully\n", username);
72        return true;
73    }
74    else
75    {
76        printf("The username or password is wrong\n");
77        return false;
78    }
79 }
80 // The function "authentication" as a vulnerable unit
81 void authentication(char *username, char *password)
82 {
83    // FLAW: Did not allocate space based on the username length
84    char *tmp_user = (char *)(malloc(80*(sizeof(char))));
85    // FLAW: Did not allocate space based on the password length
86    char *tmp_pass = (char *)(malloc(80*(sizeof(char))));
87    /* POTENTIAL FLAW:
88    data may not have enough space to hold source */
89    memcpy(tmp_user, username, strlen(username));
90    /* POTENTIAL FLAW:
91    data may not have enough space to hold source */
92    memcpy(tmp_pass, password, strlen(password));
93    int loginCnt = 0;
94    for(; loginCnt < 3; loginCnt++)
95    {
96        bool signin_res = signin(tmp_user, tmp_pass);
97        if(signin_res) { break; }
98        printf("The username or password is invalid, try again :");
99        printf("(%d from %d)\n",(loginCnt+1),3);
100       printf("Enter username : "); scanf("%s",tmp_user);
101       printf("Enter password : "); scanf("%s",tmp_pass);
102    }
103    if(loginCnt == 3) { printf("Please try later\n"); }
104 }
```

```
105 int main (int argc, char *argv[])
106 {
107     char *username = (char *)(malloc(100*(sizeof(char))));
108     char *password = (char *)(malloc(100*(sizeof(char))));
109     if(argc >= 3) { authentication(argv[1], argv[2]); }
110     else
111     {
112         printf("Register new user\n");
113         printf("Enter username :"); scanf("%s", username);
114         printf("Enter password :"); scanf("%s", password);
115         if(username[0] >= 'A' && username[0] <= 'Z')
116             signup(username, password);
117         else
118             printf("The selected username is not valid,
                    it must start with an uppercase letter");
119     }
120 }
```

Listing 2. Source code of the designed complex program

4   Conclusion and Future Works

While symbolic execution is sound and complete in theory, this method faces challenges in testing real-world programs, such as path explosion. The number of symbolic execution states may be exponential, and the whole program may not be analyzed thoroughly.

We proposed a method for applying symbolic execution technique to detect heap overflow vulnerability in executable codes. In this method, we limit the scope of symbolic execution to the test units. We also presented a method for determining the test units in the program according to the specification of heap overflow vulnerability in executable codes. In this method, we generate appropriate unit input data for detecting heap overflow according to the path and vulnerability constraints in vulnerable statements of the test unit. Then, we use machine learning techniques to estimate the relation between system and unit input data as a function and find consistent system input data that enters into the program from the beginning, causes execution of vulnerable statements in the test unit, and reveals heap overflow vulnerability. The experiments showed that this method achieves more efficient and accurate results in detecting vulnerabilities in complex programs compared to similar tools.

In the future, we are going to extend our solution to detect other vulnerability classes in executable codes, such as stack-based buffer overflow, use after free, and double free. We have to specify these vulnerabilities so that the implemented solution can automatically identify test units based on them. We also have to revise the *Cover* algorithm to calculate the vulnerability constraints based on the specified vulnerabilities and generate appropriate test data to detect them in the programs. Additionally, we are going to study other machine learning techniques for estimating the program behavior to improve the efficiency of the *ComputeMap* algorithm.

References

1. Heap Overflow Detection Tool. https://github.com/SoftwareSecurityLab/Heap-Overflow-Detection
2. National Institute of Standards and Technology in Software Assurance Reference Dataset Project. https://samate.nist.gov/SRD. Accessed 4 Mar 2022
3. Arlinghaus, S.L., Arlinghaus, W.C., Drake, W.D., Nystuen, J.D.: Practical Handbook of Curve Fitting (1994)
4. Baldoni, R., Coppa, E., D'elia, D.C., Demetrescu, C., Finocchi, I.: A survey of symbolic execution techniques. ACM Comput. Surv. **51**(3) (2018). https://doi.org/10.1145/3182657
5. Cadar, C., Dunbar, D., Engler, D.: KLEE: unassisted and automatic generation of high-coverage tests for complex systems programs. In: Proceedings of the 8th USENIX Conference on Operating Systems Design and Implementation, OSDI 2008, pp. 209–224. USENIX Association (2008). https://doi.org/10.5555/1855741.1855756
6. Cha, S., Hong, S., Bak, J., Kim, J., Lee, J., Oh, H.: Enhancing dynamic symbolic execution by automatically learning search heuristics. IEEE Trans. Softw. Engi., 1 (2021). https://doi.org/10.1109/TSE.2021.3101870
7. Cha, S., Lee, S., Oh, H.: Template-guided concolic testing via online learning, pp. 408–418. Association for Computing Machinery, New York (2018). https://doi.org/10.1145/3238147.3238227
8. Cha, S., Oh, H.: Concolic testing with adaptively changing search heuristics. In: Proceedings of the 2019 27th ACM Joint Meeting on European Software Engineering Conference and Symposium on the Foundations of Software Engineering, ESEC/FSE 2019, pp. 235–245. Association for Computing Machinery, New York (2019). https://doi.org/10.1145/3338906.3338964
9. Chen, J., Hu, W., Zhang, L., Hao, D., Khurshid, S., Zhang, L.: Learning to accelerate symbolic execution via code transformation. In: Millstein, T. (ed.) 32nd European Conference on Object-Oriented Programming (ECOOP 2018). Leibniz International Proceedings in Informatics (LIPIcs), vol. 109, pp. 6:1–6:27. Schloss Dagstuhl-Leibniz-Zentrum fuer Informatik, Dagstuhl (2018). https://doi.org/10.4230/LIPIcs.ECOOP.2018.6
10. Chen, T., Zhang, X.S., Guo, S.Z., Li, H.Y., Wu, Y.: State of the art: dynamic symbolic execution for automated test generation. Future Gener. Comput. Syst. **29**(7), 1758–1773 (2013). https://doi.org/10.1016/j.future.2012.02.006
11. Davies, M., Păsăreanu, C.S., Raman, V.: Symbolic execution enhanced system testing. In: Joshi, R., Müller, P., Podelski, A. (eds.) VSTTE 2012. LNCS, vol. 7152, pp. 294–309. Springer, Heidelberg (2012). https://doi.org/10.1007/978-3-642-27705-4_23
12. Dong, S., Olivo, O., Zhang, L., Khurshid, S.: Studying the influence of standard compiler optimizations on symbolic execution. In: 2015 IEEE 26th International Symposium on Software Reliability Engineering (ISSRE), pp. 205–215 (2015). https://doi.org/10.1109/ISSRE.2015.7381814
13. Godefroid, P.: Compositional dynamic test generation. SIGPLAN Not. **42**(1), 47–54 (2007). https://doi.org/10.1145/1190215.1190226
14. Godefroid, P., Luchaup, D.: Automatic partial loop summarization in dynamic test generation. In: Proceedings of the 2011 International Symposium on Software Testing and Analysis, ISSTA 2011, pp. 23–33. Association for Computing Machinery, New York (2011). https://doi.org/10.1145/2001420.2001424

15. Hansen, T., Schachte, P., Søndergaard, H.: State joining and splitting for the symbolic execution of binaries. In: Bensalem, S., Peled, D.A. (eds.) RV 2009. LNCS, vol. 5779, pp. 76–92. Springer, Heidelberg (2009). https://doi.org/10.1007/978-3-642-04694-0_6

16. Menzies, T., Hu, Y.: Data mining for very busy people. Computer **36**(11), 22–29 (2003). https://doi.org/10.1109/MC.2003.1244531

17. Mouzarani, M., Sadeghiyan, B.: Towards designing an extendable vulnerability detection method for executable codes. Inf. Softw. Technol. **80**, 231–244 (2016). https://doi.org/10.1016/j.infsof.2016.09.004

18. Ognawala, S., Ochoa, M., Pretschner, A., Limmer, T.: MACKE: compositional analysis of low-level vulnerabilities with symbolic execution. In: Proceedings of the 31st IEEE/ACM International Conference on Automated Software Engineering, ASE 2016, pp. 780–785. Association for Computing Machinery, New York (2016). https://doi.org/10.1145/2970276.2970281

19. Păsăreanu, C.S., et al.: Combining unit-level symbolic execution and system-level concrete execution for testing Nasa software. ISSTA 2008, pp. 15–26. Association for Computing Machinery, New York (2008). https://doi.org/10.1145/1390630.1390635

20. Schwartz-Narbonne, D., Schäf, M., Jovanović, D., Rümmer, P., Wies, T.: Conflict-directed graph coverage. In: Havelund, K., Holzmann, G., Joshi, R. (eds.) NFM 2015. LNCS, vol. 9058, pp. 327–342. Springer, Cham (2015). https://doi.org/10.1007/978-3-319-17524-9_23

21. Stephens, N., et al.: Driller: augmenting fuzzing through selective symbolic execution. In: In: NDSS (2016). https://doi.org/10.14722/ndss.2016.23368

22. Strang, G.: Linear Algebra and Its Applications. Thomson, Brooks/Cole, Belmont (2006). http://www.amazon.com/Linear-Algebra-Its-Applications-Edition/dp/0030105676

23. Wang, F., Shoshitaishvili, Y.: Angr - the next generation of binary analysis. In: 2017 IEEE Cybersecurity Development (SecDev), pp. 8–9 (2017). https://doi.org/10.1109/SecDev.2017.14

Conformance Testing of Formal Semantics Using Grammar-Based Fuzzing

Diego Marmsoler[ID] and Achim D. Brucker[(⊠)][ID]

University of Exeter, Exeter, UK
{d.marmsoler,a.brucker}@exeter.ac.uk

Abstract. A common problem in verification is to ensure that the formal specification models the real-world system, i.e., the implementation, faithfully. Testing is a technique that can help to bridge the gap between a formal specification and its implementation.

Fuzzing in general and grammar-based fuzzing in particular are successfully used for finding bugs in implementations. Traditional fuzzing applications rely on an implicit test specification that informally can be described as "the program under test does not crash".

In this paper, we present an approach using grammar-based fuzzing to ensure the conformance of a formal specification, namely the formal semantics of the Solidity Programming language, to a real-world implementation. For this, we derive an executable test-oracle from the formal semantics of Solidity in Isabelle/HOL. The derived test oracle is used during the fuzzing of the implementation to validate that the formal semantics and the implementation are in conformance.

Keywords: Conformance testing · Fuzzing · Verification · Solidity

1 Introduction

Formal verification is an important means for ensuring that systems are correct, safe, and secure. But any formal verification of a computer program can only be as good as the formal semantics provided for the corresponding programming language. If the formal semantics does not capture the behavior of the real system adequately, the results of the verification can hardly be trusted. Consequently, the conformance of a formal semantics to the actual implementation is crucial in using formal verification to build correct, safe, and secure systems.

Sadly, in many important application areas, systems are not developed with a formal semantics as a point of departure. The formal semantics is often an afterthought, developed by reverse engineering informal documentation or, more often, the behavior of real systems. Thus, the problem of understanding to what extent such a formal semantics captures the behavior of the real system adequately becomes a big challenge.

Testing is a methodology that can help us to understand the relation between a formal semantics and its implementation. For example, one can derive test cases

L. Kovács and K. Meinke (Eds.): TAP 2022, LNCS 13361, pp. 106–125, 2022.
https://doi.org/10.1007/978-3-031-09827-7_7

from a test specification expressed in a formal semantics (e.g., using theorem-prover-based testing as discussed in [9]) that are then executed and validated on the real system to test that the real system conforms to the formal specification. Another approach is using property-based testing on the formal semantics (e.g., using [10]) itself to animate and explore the formal specification.

In this paper, we present a new approach based on grammar-based fuzzing, a technique that generates example programs from a formal grammar. Our goal is to ensure that our post-hoc developed formalization complies to the real world system, i.e., our formalization should behave identical to the implementation. To this end, we use grammar-based fuzzing to generate Solidity programs that are then executed on the actual system and on the formal semantics to compare their outcomes.

Our case-study is based on a formal semantics of Solidity [36]. Solidity is a programming language for expressing smart contracts (SCs) on the Ethereum blockchain. It is a Turing-complete, statically typed programming language whose concrete syntax has been designed to look familiar to people knowing Java, C, or JavaScript. The following shows a simple (artificial) function of a SC in Solidity for a withdrawal operation:

```
1  function wd(uint256 n, address payable r) public returns(bool) {
2      if (n < address(this).balance) {
3          r.transfer(n);
4          return true;
5      }
6      return false;
7  }
```

While Solidity is in many aspects similar to, e.g., Java, it differs in others. For example, the type system of Solidity provides, e.g., numerous integer types of different sizes (e.g., uint256) and the Solidity programs can make use of different types of stores for data (e.g., storage and memory).

In more detail, our contributions are:

1. An approach extending a parse grammar for Solidity to ensure that a generic grammar-based fuzzer generates *type correct* Solidity programs (Sect. 2.2), instead of generating syntactically correct, but often ill-typed, programs.
2. An approach for automatically deriving a test-oracle from a formal specification in Isabelle/HOL that allows to efficiently decide if a test case passes or fails *and* that allows to measure the test coverage in terms of statements and expressions of the target language usually based on an implicit test specification that informally can be described as "no crashes occur".
3. A framework for testing the compliance of a formal semantics of Solidity in Isabelle/HOL to their execution on the Ethereum blockchain (Sect. 2.3).

We evaluate our approach using a denotational semantics, in Isabelle/HOL [35], for a subset of Solidity v0.5.16 [36][1] that we described in [32]. Our formal

---

[1] This is the currently supported default version of the Truffle test framework.

semantics of Solidity [32] and the implementation of our compliance testing approach is publicly available [33].

Our results (discussed in Sect. 3) suggest that the approach has great potential to detect deviations of a semantics from a reference implementation. In particular, we successfully used the framework to uncover more than 30 such deviations in our original version of the semantics (some of these deviations are discussed in Sect. 3.1).

2   Approach

Figure 1 provides a high-level overview of our approach. Its main components are:

- a formal semantics of a subset of Solidity v0.5.16 [36] in Isabelle/HOL [32],
- a test oracle that we generate automatically form the formal semantics, and
- a grammar-based fuzzer (based on Grammarinator [23]) that uses an extended (enriched with additional type information) Grammar of Solidity for generating type correct Solidity programs (i.e., test cases).

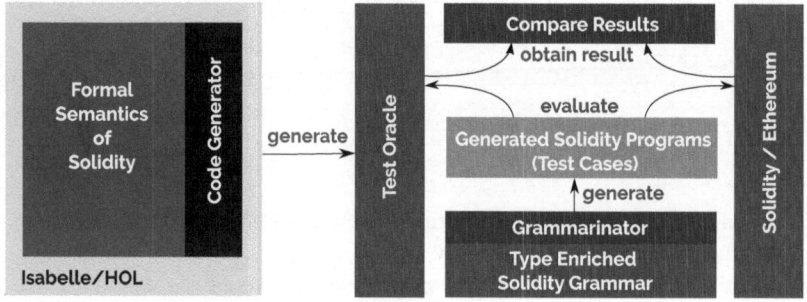

Fig. 1. Conformance testing by combining test and proof

The overall workflow is as follows: from the formal semantics of Solidity, given as a deep embedding into Isabelle/HOL, we generate a test oracle as executable command line program (see Sect. 2.1). Moreover, from a formal grammar describing the concrete syntax of Solidty, enriched with information to capture a subset of the typing rules of Solidity (see Sect. 2.2), we generated Solidity programs (i.e., test cases) using a grammar-based fuzzer (i.e., a fuzzing test tool that generates test cases systematically from a grammar, ensuring that all test cases are synatically correct with respect to the provided grammar). The results of these two steps feed into the actual test execution: the generated programs (together with their initial program state) are both executed on the Ethereum blockchain (i.e., the implementation) and our formally derived test oracle. Finally, the resulting programs states are compared and if they are equal, the test has passed, otherwise the test failed (see Sect. 2.3). We conclude this section by illustrating our approach by discussing an example SC generated by our framework (Sect. 2.4).

2.1   Deriving a Test Oracle from a Formal Semantics

We base our current work on a formalization of the Solidity in Isabelle/HOL [35], for details, we refer the reader elsewhere [32]. The core of the formal semantics is the semantic function for statements:

$$\mathcal{C}: \ \mathbf{C} \rightarrow \mathbf{Environment} \rightarrow \mathbf{State} \rightarrow \mathbf{Nat} \rightarrow (\mathbf{State} \times \mathbf{Nat})_\perp$$

where **C** is the data type capturing the statements of Solidity (e.g., while loops, conditionals, assignments), **Environment** is the environment in which the expression that are part of a statement are evaluated in, and **State** is the program state. So far, this is mostly the standard definition of a semantic function for a denotational semantics. There are a few exceptions though, in particular: the execution of Solidity statements generates Gas costs. The initial balance is passed as **Nat** and the semantic function returns a tuple consisting of the new program state and the updated Gas balance. If an execution runs out of Gas during the execution, the program is terminated (see [32,41] for details).

Assuming that the semantics function is executable, in principle, the semantics function can be used as test oracle. In its simplest form, we can use the simplifier of Isabelle to symbolically evaluate the semantic function, in Isabelle. Given a concrete Solidity program (i.e., a statement), a ground environment, state, and a concrete Gas balance, the simplification in Isabelle will yield a state (or none, in case of an exception during the program evaluation).

This straightforward approach is not always desired. Firstly, symbolic execution using Isabelle's simplifier can be very slow. Second, it requires to interface the test framework directly with Isabelle. Thus, we make use of Isabelle's code generator to generate a Haskell implementation of the semantic function that can be compiled into a stand-alone program (i.e., the Test Oracle in Fig. 1). This program takes the same arguments as our semantic evaluation function and returns either an error or a new program state, which is used for determining if our semantics, for a given test case (i.e., a Solidity program and corresponding state) is compliant to the implementation.

Naturally, a formal semantics expressed in higher-order logic (HOL) is more abstract than a program in a programming language, such as Haskell [31]. For example, our semantics makes intensive use of (finite) sets, which results in a semantic function that is not executable in a strict sense. To make it efficiently executable, we need to transform expressions using sets into semantically equivalent expressions using lists. For this, we formally prove in Isabelle conversion lemmas, e.g.:

- **lemma [code]:**
  sorted_list_of_set (set xs) = sort (remdups xs)
- **lemma [code]:**
  ffold_init ct a c = fold (init ct) (remdups (sorted_list_of_set (fset c))) a

These lemmas, proved for improving our testing approach, are added to the simplifier used for generating code, i.e., automatically converting sets to a more efficient list representation.

Moreover, we need to interface our test oracle with the test system, i.e., our generated program needs to be able to parse (the abstract syntax of) Solidity programs. For this reason, we have chosen Haskell as the target language (instead of one of the other targets Isabelle's code generator supports): Haskell allows us to make use of the automated generation of parsing and pretty-printing functions using Haskell's *deriving* feature.

Out setup is also used by Isabelle's code generator for SML, the language Isabelle's kernel is implemented in. This makes fast and efficient evaluation of Solidity expressions available to Isabelle itself, e.g., the value-statement:

```
value eval 1 stmt SKIP (STR ''089B'') (STR '''') (STR ''0'')
      [(STR ''089'', STR ''100''),(STR ''15f'', STR ''100'')]
      [] [(STR ''v1'', (Value TBool, Stackbool True))]
```

yields `"STR ''v1==true\n089.balance==100\n15f.balance==100\n''"`. This setup is also automatically used by (potentially unsafe) proof tactics such as "eval" or Isabelle's specification-based testing tools and counter-example generators such as QuickCheck.

2.2   Generate Random Solidity Code

Grammarinator [23] is a grammar-based fuzzing tool: given a formal grammar of a language, Grammarinator generates random programs, which are syntactically correct (but, e.g., could still be ill-typed). The quality of the generated programs, however, depends on the provided grammar. In particular, if the grammar is too relaxed, this leads to the creation of programs which do not even compile. For the purpose of testing the semantics, however, we are interested in programs which we can run and compare with our semantics. Thus, we need to provide a grammar which is strong enough to lead to the creation of compiling programs.

In general, such a grammar needs to consider typing information and thus consists of more rules as we usually see in grammars used for the generation of parsers. For example, the grammar we used to test our subset of Solidity consists of more than 35 000 lines of ANTLR4 code. Thus, instead of manually creating the grammar, we implemented a tool which generates the corresponding grammar. Roughly speaking, the program consists of six main steps: 1. Generate rules for types 2. Generate rules for identifiers 3. Generate rules for variable declarations 4. Generate rules for expressions 5. Generate rules for lvalues 6. Generate rules for statements In the following, we discuss the rules generated by the tool.

Rules for Types. Our subset of Solidity supports four basic types: boolean, address, signed and unsigned integers of 8–256 bit. Thus, the generated grammar contains rules for all of these types.

Rules for Identifiers. In Solidity, we usually distinguish between *basic types* and *complex types* such as mappings and arrays. In addition, Solidity supports two different kinds of store to keep corresponding values: *memory* and *storage*. Our grammar provides rules to generate identifiers for all of these types.

In particular, the grammar provides one rule for producing identifiers for each of the basic types. Each of these rules allows for the creation of up to 10 identifiers for each kind of store:

```
IB:  'v_b_' ('s'|'m') [0-9] ;
IA : 'v_a_' ('s'|'m') [0-9] ;
IU8: 'v_u8_' ('s'|'m') [0-9] ; ...
IS8: 'v_s8_' ('s'|'m') [0-9] ; ...
```

In Solidity, mappings are only allowed to be kept in storage which is why we do not need to distinguish between memory and storage for mapping identifiers. The type of a mapping depends on the type of its keys and the type of its values, and thus we need to consider both of them when generating corresponding identifiers. In the following we show one such rule used to generate identifiers for mappings from addresses to unsigned 128-bit integers:

```
IMAU128: 'v_m_a_u128_' [0-9] ;
```

Similar rules are used to generate identifiers for all other types of mappings.

Array types are determined by the type of their values and the size of each of their dimensions. Thus, we need to consider all of them when generating rules for the production of array identifiers. We now show two such rules used to generate identifiers for a two-dimensional storage/memory array of signed 88-bit integers in which the first dimension is of size 3 and the second one of size 2:

```
IAS88_S_32: 'a_s88_32_s' [0-9] ;
IAS88_M_32: 'a_s88_32_m' [0-9] ;
```

Again, similar rules are used to generate identifiers for all other types of arrays.

Rules for Variable Declarations. In Solidity, mappings can only be used at the SC level and not for local declarations. Thus, the rules for local declarations only need to consider variables of basic types as well as variables of array type. However, the rules need to ensure consistency of the identifier with its type. For base types we only need to combine the corresponding rules for identifiers and types discussed above.

For arrays, however, the situation is more complicated. First, a declaration needs to explicitly state the store in which the array is kept (storage or memory). In addition, Solidity does not allow for uninitialized storage arrays. These two aspects need to be considered when creating rules for the generation of variable declarations. In the following we list two of these rules used to generate variable declarations for a two-dimensional storage/memory array of signed 88-bit integers in which the first dimension is of size 3 and the second one of size 2:

```
TS88 '[3][2]' ' ' STORAGE ' ' IAS88_S_32
  '=' as88_S_32_exp ';'
TS88 '[3][2]' ' ' MEMORY ' ' IAS88_M_32
  ('=' (as88_S_32_exp | as88_M_32_exp))? ';'
```

Note that initialization is required for the version dealing with storage variables and optional for the version dealing with memory variables. In addition, in the first case, the variable needs to be initialized with a corresponding storage expression while in the latter case the variable may be initialized with either a storage or a memory expression as long as the types are consistent.

Rules for Expressions. Our grammar provides rules to generate expressions for each of the different base types. In particular, expressions of a certain type can be directly generated by literals or identifiers of that type. In addition, an expression of a certain type can be obtained from identifiers of higher types by using corresponding keys. For example, the following rules are used to produce boolean expressions from corresponding mapping identifiers:

```
| IMBB '[' bexpression ']'
| IMAB '[' aexpression ']'
| IMU8B '[' u8expression ']' ...
| IMS8B '[' s8expression ']' ...
```

Again, the situation is more complicated when it comes to array identifiers. To this end, we first need to add rules to generate corresponding keys:

```
I1: 'uint256(' [0-0] ')' ;
I2: 'uint256(' [0-1] ')' ;
I3: 'uint256(' [0-2] ')' ;
```

Here, I_k is a rule to access one dimension of an array of size k. Now we can use these rules to create rules to access the various dimensions of an array. For example, the following rules are used to generate an expression of a one-dimensional boolean memory array of size 3:

```
ab_M_3_exp                          ab_M_31_exp : IAB_M_31 ;
  : IAB_M_3
  | ab_M_31_exp '[' I1 ']'          ab_M_32_exp : IAB_M_32 ;
  | ab_M_32_exp '[' I2 ']'
  | ab_M_33_exp '[' I3 ']'          ab_M_33_exp : IAB_M_33 ;
  ;
```

In particular, such an expression can be obtained by either using a corresponding identifier or by an expression of a two-dimensional memory boolean array using a corresponding key value. Finally, we can use the rules for array expressions to create rules for the corresponding base type. For example, the following rules are used to produce boolean expressions from corresponding array identifiers:

```
| ab_S_1_exp '[' I1 ']' ...
| ab_M_1_exp '[' I1 ']' ...
```

Having rules for base expressions of a certain type we can then combine them using corresponding operators. For example, the following rules are used to combine boolean expressions using logical operators:

```
| bexpression OP_EQ bexpression
| bexpression (OP_AND|OP_OR) bexpression
```

Care needs to be taken if we use operators over integers since in Solidity only certain types of integers may be combined. The following rules apply:

- Signed integers can be compared to other signed integers.
- Unsigned integers can be compared to other unsigned integers.
- Signed integers can only be compared to unsigned integers with smaller size.

For example, the following rule generates boolean expressions by comparing an unsigned 16-bit integer with a signed 24-bit integer:

```
u16expression (OP_EQ|OP_LE) s24expression
```

However, our grammar does not provide rules to compare, for example a signed 16-bit integer with an unsigned 24 bit one since these two types are not compatible and thus the corresponding program would not compile.

The rules for the generation of lvalues is similar to the rules for expressions and not discussed further here.

Rules for Statements. In general, our grammar provides two types of rules to generate basic Solidity statements: *assignments* and *transfer* commands.

Again, we need to be careful when creating assignment rules since they may easily lead to type errors and thus programs which would not compile. In particular, we need to ensure that lvalues and corresponding expressions have compatible types. This is again simple for expression of basic types. However, the situation is again more complex when it comes to arrays because we need to consider the dimension on the involved arrays. For example, the following rules are used to generate assignments for two-dimensional storage/memory arrays of size 3 and 2:

```
| ab_S_32_lval '=' (ab_S_32_exp | ab_M_32_exp) ';'
| ab_M_32_lval '=' (ab_S_32_exp | ab_M_32_exp) ';'
```

Note that Solidity does indeed allow the assignment of arrays located in different stores as long as the dimensions and types are compatible.

The transfer command allows transferring money from one account to another and our grammar provides corresponding rules for 10 different accounts:

```
| AC0 '.' TRANSFER '(' u8expression ')' ';'
| AC1 '.' TRANSFER '(' u8expression ')' ';'
| AC2 '.' TRANSFER '(' u8expression ')' ';'      ...
```

Note that we only allow for unsigned integers of 8-bit expressions to be used in transfers to avoid that balances become empty too fast.

Finally, we can provide rules for higher-level statements such as blocks, conditionals, and loops which completes our grammar:

```
| '{' declaration statement* '}'
| 'if' '(' bexpression ')' '{' statement '}'
    'else' '{' statement '}'
| 'while' '(' bexpression ')' '{' statement '}'
```

2.3   Testing Algorithm

The test framework is fully automated. Algorithm 1 shows the core algorithm, where [] denotes the empty list and $xs \leftarrow^{+} x$ denotes the list resulting from appending element x to list xs. The algorithm requires three configurations:

noStmt the number of programs we would like to test
noStates the number of states for each program which we would like to test
grammar the Solidity grammar file used to generate Solidity programs

It will then execute noStmt×noStates tests and return a list of failing statements with corresponding states.

The algorithm proceeds in noStmt rounds (line 2–line 20). For each round i, it first generates a random Solidity program stmt (line 3). It uses the grammar-based fuzzer Grammarinator [23] and the Solidity grammar grammar discussed in Sect. 2.2. Occasionally, the fuzzer may generate a statement twice, which is why we need to check if the statement was already processed (line 4–line 5).

Next, we analyze the generated program and extract all the variable identifiers vars (line 6). Note that this step requires to be able to identify an identifier

---

Algorithm 1: TestSolidity

Data: noStmt
Data: noStates
Data: grammar
Result: results containing statements which lead to different results

```
 1  results, stmts ← [];
 2  for i ← 0 to noStmt do
 3  │   stmt ← generate(grammar);
 4  │   if stmt ∈ stmts then
 5  │   │   continue;
 6  │   vars ← extract(stmt);
 7  │   istmt ← instrument(stmt);
 8  │   astmt ← parse(istmt);
 9  │   scs ← [];
10  │   for i ← 0 to noStates do
11  │   │   vals ← [];
12  │   │   for var ∈ vars do
13  │   │   │   val ← random(var);
14  │   │   │   vals ←⁺ var, val;
15  │   │   result ← evaluate(astmt,vals);
16  │   │   scs ←⁺ createSC(vals,istmt,result);
17  │   result ← execute(scs);
18  │   if ¬result then
19  │   │   results ←⁺ astmt, vals;
20  │   stmts ←⁺ stmt;
```

within program code. However, this is possible since, as discussed in Sect. 2.2, our grammar requires a certain structure for variable identifiers. Indeed, our grammar even ensures that the type of an identifier is encoded in its name which is required later on for generating random states.

The generated programs may also contain loops, and thus they may not always terminate. However, for the purpose of testing the implementation against the semantics, termination is actually desirable. Thus, line 7 analyzes the generated program and modifies all loops by adding a variable which increases with each iteration. In addition, it adds a disjunction to the loop condition which asserts that the variable is less than a certain value (see Sect. 2.4 for an example).

The resulting program istmt is in concrete Solidity syntax. However, our semantics requires a program given in abstract syntax. Thus, line 8 parses the modified program and returns its abstract syntax tree astmt.

Next we generate noStates random states for the currently processed program (line 10–line 16). To this end, we first iterate over all the variables extracted from the statement, generate a random value for it and add the variable as well as its value to a list vals (line 12–line 14). Note that for a meaningful program, the generated value needs to conform to the type of the identifier. However, as mentioned above, the type of an identifier is encoded into its name which allows for the generation of type-conform values.

Having generated a random state, we can execute the abstract statement using our evaluator. The resulting state is then used to create a corresponding test SC and add it to list scs. To this end, we modify a template SC with a single test function: The variables and their generated values vals are used to create corresponding variable declarations. The body of the function is given by the concrete statement istmt. The result state result from the evaluator is used to create corresponding assertions. An example of such an SC is discussed later on in Sect. 2.4.

After creating test SCs for each state we can execute them using the Truffle test framework [14]. The framework deploys the SCs to a local instance of the Ganache blockchain [13] and executes their test methods reporting failing assertions. Note that if an assertion fails, this means that the computed results for one of the variables deviates from the corresponding value of the result state obtained by the evaluator, and thus it will be recorded by the algorithm (line 18–line 19) in list results.

Finally, the statement is added to the list of processed statements stmts and a new iteration starts.

2.4 Example Smart Contract

Listing 1.1 shows parts of an SC generated from our test framework. As mentioned in Sect. 2.3, the SC consists of a single method test which contains the generated program. The program itself (istmt in Algorithm 1) is provided in line 13–line 20. Note that the program contains a while loop, and thus it was modified by our instrumentation (method `instrument` in Algorithm 1) which added a counter variable `counter1` to ensure termination.

The extracted storage variables are declared as SC variables (line 2–line 5) whereas the extracted memory/stack variables are declared locally (line 7–line 9). The generated state (`vals` in Algorithm 1) was used to initialize the variables extracted from the statement (line 10–line 12).

Finally, the outcome of the evaluator (`result` in Algorithm 1) was used to generate assert statements (line 21–line 27).

3   Evaluation

Our framework detected over 30 issues with our original semantics. In the following we discuss some of them and then present statistics about test execution.

```
1    contract TestContract0 {
2        uint8 v_u8_s8;
3        mapping(uint16 => uint8) v_m_u16_u8_9;          Extracted
4        bool[1][2] a_b_12_s5;                           storage variables
5        ...
6        function test() public {
7            uint104 v_u104_m2;                          Extracted
8            uint104[1][1] memory a_u104_11_m2;          memory/stack variables
9            ...
10           v_u104_m2=14622709355569675963178665339646;   Generated
11           v_m_u16_u8_9[59381]=79;                        input state
12           ...
13           int8 counter1=int8(0);
14           while((v_m_u224_s240_1[uint224(444)]==
15               (v_u216_s1-v_u104_m2)) && counter1<int8(10)){
16               0xf7218C33533a3F22e3296F8b1DC0074B399355Eb     Generated
17                   .transfer(v_m_u16_u8_9[uint16(0)]);        program
18               counter1=counter1+int8(1);
19           }
20           ...
21           Assert.equal(v_m_u16_u8_9[59381]==79, true);
22           Assert.equal(a_u104_11_m2[0][0]==
23               813009781905416963279596089600  7, true);
24           Assert.equal(                                   Computed
25               0xf7218C33533a3F22e3296F8b1DC0074B399355Eb   result state
26               .balance==100000000000000000000, true);
27           ...
28       }
29   }
```

Listing 1.1. Example test contract generated by our testing framework.

3.1 Examples of Detected Issues

Integer Arithmetic. In Solidity, arithmetic operations are allowed if the types of the operators are compatible (see Sect. 2.2). Consider, for example, the following Solidity statements:

```
assert (uint128(1) + int256(1) == int256(2));
assert (uint128(1) + uint256(1) == uint256(2));
assert (int128(1) + int256(1) == int256(2));
// uint256(1) + int128(1) is not allowed
```

As can be seen, arithmetic operations are possible if both operands are either signed or unsigned integers or if one is signed and the other unsigned and the size of the signed one is larger than the size of the unsigned one.

In the original version of our semantics we followed this rule. However, due to a misplaced comparison operator we assigned an error to expressions in which the first operand was an *unsigned* b_1-bit integer and the second operand a *signed* b_2-bit integer and $b_1 < b_2$.

Implicit Initialization. In Solidity, uninitialized variables are implicitly assigned a default value. Consider, for example, the following Solidity fragments:

```
int128 x;            bool x;              address x;
assert(x==0);        assert(x==false);    assert(x==address(0x0));
```

Here, a variable of type *integer* is implicitly initialized with 0, a variable of type *bool* with `false`, and a variable of type address with address 0.

We followed this specification in our original semantics, however, due to a mistake in the initialization function, signed integer variables were initialized with their size instead of 0.

Storage Pointers. In Solidity, assignments between variables denoting storage arrays copy the array from one storage location to another. Consider, for example, the smart contract (SC) Test1 depicted in Fig. 2. Here, the assignment in line 3 actually copies the whole array from var1 to var2. Thus, the assignment in line 6 only affects the copy in var2 and var1 remains unchanged.

This behavior changes, however, if we use a local storage variable. To see this, consider SC Test2 in Fig. 2. Here, the assignment in line 5 only assigns the storage location of var1 to var2. Thus, the assignment in line 6 actually changes the value of var1. In our original semantics we did not make this distinction and rather always copied the complete array.

Array Copy. In Solidity, assignments between variables of arrays within different stores require the arrays to be copied between the stores. Consider, for example, the following SC:

```
1      contract Test {
2          uint8[2][2] var1=[[1,1], [1,1]];
```

```
1  contract Test1 {                      1  contract Test2 {
2     uint8[2] var1=[1,2];               2     uint8[2] var1=[1,2];
3     uint8[2] var2=var1;                3
4                                        4     function test() public {
5     function test() public {          5        uint8[2] storage var2=var1;
6        var2[1]=0;                      6        var2[1]=0;
7        assert(var1[1]==1);             7        assert(var1[1]==0);
8     }                                  8     }
9  }                                     9  }
```

Fig. 2. Example solidity smart contract.

```
3
4        function test() public {
5           uint8[2] memory var2 = [2,2];
6           var1[1]=var2;
7           assert(var1[0][0]==1);
8           assert(var1[0][1]==1);
9           assert(var1[1][0]==2);
10          assert(var1[1][1]==2);
11       }
12    }
```

Here, the array var1 is located in storage whereas the array var2 is located in memory. Thus, the assignment in line 6 requires the array of var2 to be copied to the location of the second array in var1. This changes all entries in var1 where the first index is 1.

In our original semantics we considered the requirement to copy arrays in assignments when they are located in different stores. However, we made a mistake in calculating the storage locations: in our original version of the semantics, we would have changed all entries in var1 where the second index is 1.

3.2 Statistics

After fixing all the detected bugs, we run the framework for several days which resulted in more than 10 000 successful tests. To cross-validate the effectiveness of the testing framework we also collected coverage information for the semantics using the Hpc tool [19]. The results are summarized in Fig. 3: Out of 123 definitions, 121 were executed during the tests. In addition, 186 alternatives (out of 524) and 1 592 expressions (out of 2 394) were executed.

Hpc also generates detailed coverage reports for every module. When inspecting these reports it turns out that the low number of covered alternatives is mainly because of missing executions of error cases (e.g. ill-typed programs). Consider, for example, Fig. 4 which shows an excerpt of the coverage report for the semantics of conditionals. From the figure we can observe that most of the code is indeed executed. In particular, the fuzzer generated programs and corresponding states which triggered the execution of both: the true and the false

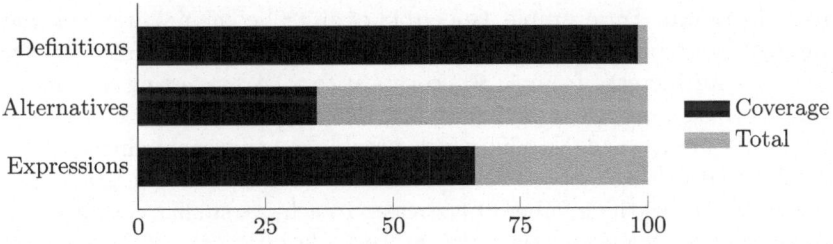

Fig. 3. Overall test coverage of semantics.

branch. In addition, we can see that the only code which was not executed is the one dealing with erroneous situations such as non-compiling programs.

```
208  stmt bal (ITE ex s1 s2) e st =
209    let {
210      gas = costs_min (ITE ex s1 s2) e st;
211    } in (if Arith.less_eq_nat bal gas then (Out_of_Gas, gas)
212         else (case Expressions.expr ex e st of {
213           Nothing -> (Error, gas);
214           Just (Storage.KValue b, Environment.Value Valuetypes.TBool) ->
215             (case (if b == ReadShow.showL_b_o_o_1 True
216                  then stmt (Arith.minus_nat bal gas) s1 e st
217                  else stmt (Arith.minus_nat bal gas) s2 e st)
218               of {
219               (res, c) -> (res, Arith.plus_nat c gas);
220             });
221           }));
```

Fig. 4. Coverage report for conditional from `Statements.hs.html` [33]

This is because our framework only generates well-formed Solidity programs and thus the error cases are not executed. While this significantly increases the amount of good test cases it can also be a limitation if deviations occur in error cases. In particular, the framework does not allow to test that a program which does not compile leads to an error in the semantics. While this may be acceptable in some situations, it may not be acceptable for other situations which is why we are currently working on this as mentioned in future work.

4   Related Work

Grammar-Based Fuzzing. Fuzzing is a common technique to automatic testing [17]. Grammar-based testing is one technique for fuzzing structured inputs. First work in this area dates back to 1970s [22,38]. Modern tools in this area comprise, for example Grammarinator [23] and LangFuzz [24]. A stochastic approach is provided by Kifetew et al. [28].

The problem with these types of approaches is that traditional grammars are too relaxed (recall Sect. 2.2) and thus create only few relevant inputs. This problem with grammar-based fuzzing is well-known and attempts have been made to improve the results. For example, Godefroid et al. [20] combines grammar-based fuzzing with whitebox testing to increase the number of meaningful tests. Thus, they achieve an increase from 11.9% coverage to 20% for testing a JavaScript interpreter using grammar-based whitebox fuzzing. Another work in this area is due to Majumdar and Xu [30] which provide an approach which combines grammar-based testing with concolic execution based on symbolic grammars to significantly reduces the number of test cases to achieve similar coverage.

While all these works are related to our work, the objectives of the approaches are different. Work in this area usually generates test input to look for runtime errors whereas with our work we want to detect semantic deviations between a semantic specification and a reference implementation.

Validation of Semantics. According to Blazy and Leroy [7], there are five basic methods to validate formal semantics:

M1 Manual review and debugging
M2 Proving properties of the semantics, such as type preservation and determinism
M3 Using verified translations and trusted semantics
M4 Validating executable semantics, e.g. testing against test suites and experimental testing
M5 Using equivalent, alternate versions of the semantics

For example, many of the current available semantics for Solidity [2,5,15,34] are validated using M2.

If the semantics is executable then M4 is a common approach. For example, [8] validates the formal semantics of the Document Object Model (DOM), in Isabelle/HOL, by symbolically executing test cases from the official compliance test suite. Similarly, Filaretti and Maffeis [18], provide a formal semantics for PHP and validate it by executing 216 tests from the PHP Zend test suite and comparing the results with the Zend Engine. Another example is Politz et al. [37] which provide an executable semantics for JavaScript and validated it by executing 11 606 tests from the ES5 conformance suite. There exist even some examples of Soldity semantics which use this approach. Jiao et al. [26], for example, provide an executable semantics for Solidity in $\mathbb{K}$ [39] and validate it by executing 464 tests from the Solidity compiler test suite [1]. Another example is due to Yang and Lei [42] which provide an executable semantics for Solidity in Coq [40]. While all these works focus on the validation of semantics, none of them employ automatic fuzzing techniques to do so.

Using Grammar Based Fuzzing for Semantic Validation. There are some examples of work which try to automate the task to validate semantics. For example, Guagliardo and Libkin [21] provide a formal semantics for SQL queries and validate it by implementing a custom query generator to generate 100 000 tests.

Most closely to our work, however, is the work of Bereczky et al. [6] where they validate formal semantics by property-based cross-testing. Here the authors describe an approach in which they use grammars to synthesize programs which they then use to compare a semantics in the $\mathbb{K}$ framework [39] to a reference implementation There are, however, two notable differences to our work:

- They use their approach for validating the semantics of Erlang [4], a functional programming language. Thus, they avoid many of the problems occurring when you use the approach for an imperative language such as Solidity. In particular, they did not need to combine the grammar based fuzzer with random state generation as we did.
- In addition, their grammar does not seem to consider typing information which usually leads to a high number of low-quality inputs.

Compiler Testing. Another relevant area of related work is the domain of compiler testing (see [11] for an overview). In particular, one could classify our work as a grammar-directed compiler testing approach with a formally verified test oracle. Our approach of enriching the input grammar with additional information to generate type-correct programs is closely related to the use of attributed translation grammars of Duncan and Hutchison [16]. One notable difference to our work, however, is that work in this area usually does not use a formally verified test oracle.

Integrating Test and Proof. Besides the works in the area of grammar-based fuzzing for semantic validation, there are several works combining test and proof. Here, we see two areas particularly closely related to our work: First, the integration of tools inspired by QuickCheck [12] into interactive theorem provers, e.g., Isabelle/HOL [10,27]. These tools are particularly valuable for finding counter-examples prior to proof attempts of stated lemmata, saving resources in proof attempts that are deemed to fail. Moreover, such tools support the validation of the internal consistency of a formal semantics. Second, there is at least one tool, namely, HOL-TestGen [9], that derives actual test cases from a formal specification given in Isabelle/HOL. Notably, HOL-TestGen does not only generate test-cases, it also generates the test oracle that is used, during test execution, for checking if a test case passed or not. Compared to our work, HOL-TestGen generates a (small) test oracle for each individual test case while, in our current work, we derive a generic test oracle covering the whole formal semantics that can be used to check arbitrary test cases.

5 Discussion

Soundness of Results. An important aspect to consider is the correctness of Algorithm 1. Most of the steps performed by the algorithm are concerned with optimizing the quality of the produced test cases and as such they are not critical to the soundness of the approach. There is, however, one step which is indeed critical to ensure soundness which is the parsing of a generated program to an abstract syntax tree. If the parser modifies the structure of the program, then this could lead to wrong test results. Thus, it is important to ensure that the parser does not modify the structure of a program.

Grammar Quality. The quality of the results produced by this approach strongly depends on the grammar which is used for the fuzzing of programs. If the grammar is too relaxed, the fuzzer will generate mostly non-compiling programs which are not very useful for testing the semantics. If the grammar is too restrictive, some types of programs will not be tested, at all. In general, the quality of the grammar can be improved by inspecting code coverage reports. If the reports indicate that certain parts of the semantics were not executed, at all, then, the grammar is probably too restrictive and needs to be adapted.

Random States. Finally, we would like to point out that as of now, program states are generated in a purely randomized fashion. While we do ensure that the values satisfy the type of a variable, there may be more efficient ways to generate states. Thus, improvements in this aspect could further increase efficiency of the test cases.

6   Conclusion

The problem that formal semantics need to be validated against their implementation is well-known, see, e.g., [3, 25, 29]. We address this problem, in this paper, by presenting an approach to validate formal semantics against a reference implementation using grammar-based fuzzing in combination with a test oracle generated from a formal semantics given in an interactive theorem prover, namely, Isabelle/HOL.

We evaluate the approach by using it to test conformance of a Solidity semantics against their reference implementation. Our results are promising in that the framework was able to uncover more than 30 deviations in the original semantics. In addition, an analysis of code coverage shows that the approach leads to high coverage results of more than 98% for top-level definitions, more than 66% for expressions, and more than 35% for alternatives. Inspecting the code coverage reports revealed that most of the code not executed deals with erroneous situations. While it is indeed desirable to keep the number of erroneous programs low, there should still be some programs creates which is why future work should investigate how to include a low percentage of erroneous programs.

Acknowledgements. We would like to thank Tobias Nipkow for useful discussions about the compliance testing. Moreover, we would like to thank Silvio Degenhardt for his support with implementing the semantics.

Availability. Our formalization, the test framework, and the evaluation results are available under BSD license (SPDX-License-Identifier: BSD-2-Clause) [33].

References

1. Solidity. https://github.com/ethereum/solidity. Accessed 29 Mar 2022
2. Ahrendt, W., Bubel, R.: Functional verification of smart contracts via strong data integrity. In: Margaria, T., Steffen, B. (eds.) ISoLA 2020. LNCS, vol. 12478, pp. 9–24. Springer, Cham (2020). https://doi.org/10.1007/978-3-030-61467-6_2
3. Feo-Arenis, S., Westphal, B., Dietsch, D., Muñiz, M., Andisha, S., Podelski, A.: Ready for testing: ensuring conformance to industrial standards through formal verification. Formal Aspects Comput. **28**(3), 499–527 (2016). https://doi.org/10.1007/s00165-016-0365-3
4. Armstrong, J.: Programming Erlang: Software for a Concurrent World. Pragmatic Bookshelf (2013)
5. Bartoletti, M., Galletta, L., Murgia, M.: A Minimal core calculus for solidity contracts. In: Pérez-Solà, C., Navarro-Arribas, G., Biryukov, A., Garcia-Alfaro, J. (eds.) DPM/CBT -2019. LNCS, vol. 11737, pp. 233–243. Springer, Cham (2019). https://doi.org/10.1007/978-3-030-31500-9_15
6. Bereczky, P., Horpácsi, D., Kőszegi, J., Szeier, S., Thompson, S.: Validating formal semantics by property-based cross-testing. In: IFL 2020: Proceedings of the 32nd Symposium on Implementation and Application of Functional Languages, IFL 2020, pp. 150–161. Association for Computing Machinery, New York (2020). https://doi.org/10.1145/3462172.3462200
7. Blazy, S., Leroy, X.: Mechanized semantics for the Clight subset of the C language. J. Autom. Reason. **43**(3), 263–288 (2009)
8. Brucker, A.D., Herzberg, M.: Formalizing (Web) standards. In: Dubois, C., Wolff, B. (eds.) TAP 2018. LNCS, vol. 10889, pp. 159–166. Springer, Cham (2018). https://doi.org/10.1007/978-3-319-92994-1_9
9. Brucker, A.D., Wolff, B.: On theorem prover-based testing. Formal Aspects Comput. **25**(5), 683–721 (2013). https://doi.org/10.1007/s00165-012-0222-y
10. Bulwahn, L.: The new quickcheck for Isabelle. In: Hawblitzel, C., Miller, D. (eds.) CPP 2012. LNCS, vol. 7679, pp. 92–108. Springer, Heidelberg (2012). https://doi.org/10.1007/978-3-642-35308-6_10
11. Chen, J., et al.: A survey of compiler testing. ACM Comput. Surv. **53**(1) (2020). https://doi.org/10.1145/3363562
12. Claessen, K., Hughes, J.: QuickCheck: a lightweight tool for random testing of Haskell programs. In: The Fifth ACM SIGPLAN International Conference on Functional Programming, pp. 268–279. ACM Press (2000). https://doi.org/10.1145/351240.351266
13. ConsenSys Software Inc.: Ganache. https://www.trufflesuite.com/docs/ganache/. Accessed 1 May 2021
14. ConsenSys Software Inc.: Truffle. https://www.trufflesuite.com/truffle. Accessed 1 May 2021
15. Crafa, S., Di Pirro, M., Zucca, E.: Is solidity solid enough? In: Bracciali, A., Clark, J., Pintore, F., Rønne, P.B., Sala, M. (eds.) FC 2019. LNCS, vol. 11599, pp. 138–153. Springer, Cham (2020). https://doi.org/10.1007/978-3-030-43725-1_11
16. Duncan, A.G., Hutchison, J.S.: Using attributed grammars to test designs and implementations. In: Proceedings of the 5th International Conference on Software Engineering, ICSE 1981, pp. 170–178. IEEE Press (1981)
17. Felderer, M., Büchler, M., Johns, M., Brucker, A.D., Breu, R., Pretschner, A.: Security testing: a survey. Adv. Comput. **101**, 1–51 (2016). https://doi.org/10.1016/bs.adcom.2015.11.003

18. Filaretti, D., Maffeis, S.: An executable formal semantics of PHP. In: Jones, R. (ed.) ECOOP 2014. LNCS, vol. 8586, pp. 567–592. Springer, Heidelberg (2014). https://doi.org/10.1007/978-3-662-44202-9_23
19. Gill, A., Runciman, C.: Haskell program coverage. In: Haskell Workshop, Haskell 2007, pp. 1–12. ACM (2007). https://doi.org/10.1145/1291201.1291203
20. Godefroid, P., Kiezun, A., Levin, M.Y.: Grammar-based whitebox fuzzing. SIGPLAN Not. **43**(6), 206–215 (2008). https://doi.org/10.1145/1379022.1375607
21. Guagliardo, P., Libkin, L.: A formal semantics of SQL queries, its validation, and applications. Proc. VLDB Endow. **11**(1), 27–39 (2017). https://doi.org/10.14778/3151113.3151116
22. Hanford, K.V.: Automatic generation of test cases. IBM Syst. J. **9**(4), 242–257 (1970)
23. Hodován, R., Kiss, A., Gyimóthy, T.: Grammarinator: a grammar-based open source fuzzer. In: Automating TEST Case Design, A-TEST 2018, pp. 45–48. ACM (2018). https://doi.org/10.1145/3278186.3278193
24. Holler, C., Herzig, K., Zeller, A.: Fuzzing with code fragments. In: 21st USENIX Security Symposium (USENIX Security 12), pp. 445–458. USENIX Association, Bellevue, August 2012
25. Horl, J., Aichernig, B.K.: Validating voice communication requirements using lightweight formal methods. IEEE Softw. **17**(3), 21–27 (2000). https://doi.org/10.1109/52.896246
26. Jiao, J., Kan, S., Lin, S.W., Sanan, D., Liu, Y., Sun, J.: Semantic understanding of smart contracts: executable operational semantics of Solidity. In: SP, pp. 1695–1712. IEEE (2020)
27. Kappelmann, K., Bulwahn, L., Willenbrink, S.: Speccheck - specification-based testing for Isabelle/ML. Arch. Formal Proofs (2021). https://isa-afp.org/entries/SpecCheck.html. Formal Proof Development
28. Kifetew, F.M., Tiella, R., Tonella, P.: Combining stochastic grammars and genetic programming for coverage testing at the system level. In: Le Goues, C., Yoo, S. (eds.) SSBSE 2014. LNCS, vol. 8636, pp. 138–152. Springer, Cham (2014). https://doi.org/10.1007/978-3-319-09940-8_10
29. Kristoffersen, F., Walter, T.: TTCN: towards a formal semantics and validation of test suites. Comput. Netw. ISDN Syst. **29**(1), 15–47 (1996). https://doi.org/10.1016/S0169-7552(96)00016-5
30. Majumdar, R., Xu, R.G.: Directed test generation using symbolic grammars. In: The 6th Joint Meeting on European Software Engineering Conference and the ACM SIGSOFT Symposium on the Foundations of Software Engineering: Companion Papers, pp. 553–556. Association for Computing Machinery, New York (2007). https://doi.org/10.1145/1295014.1295039
31. Marlow, S.: Haskell 2010 language report (2010). https://www.haskell.org/onlinereport/haskell2010/
32. Marmsoler, D., Brucker, A.D.: A denotational semantics of solidity in Isabelle/HOL. In: Calinescu, R., Păsăreanu, C.S. (eds.) SEFM 2021. LNCS, vol. 13085, pp. 403–422. Springer, Cham (2021). https://doi.org/10.1007/978-3-030-92124-8_23. https://www.brucker.ch/bibliography/abstract/marmsoler.ea-solidity-semantics-2021
33. Marmsoler, D., Brucker, A.D.: A denotational semantics of Solidity in Isabelle/HOL: implementation and test data (2021). https://doi.org/10.5281/zenodo.5573225

34. Mavridou, A., Laszka, A., Stachtiari, E., Dubey, A.: VeriSolid: correct-by-design smart contracts for Ethereum. In: Goldberg, I., Moore, T. (eds.) FC 2019. LNCS, vol. 11598, pp. 446–465. Springer, Cham (2019). https://doi.org/10.1007/978-3-030-32101-7_27

35. Nipkow, T., Wenzel, M., Paulson, L.C. (eds.): Isabelle/HOL. LNCS, vol. 2283. Springer, Heidelberg (2002). https://doi.org/10.1007/3-540-45949-9

36. Online: Solidity documentation. https://docs.soliditylang.org/en/v0.5.16/. Accessed 1 May 2021

37. Politz, J.G., Carroll, M.J., Lerner, B.S., Pombrio, J., Krishnamurthi, S.: A tested semantics for getters, setters, and eval in JavaScript. In: Proceedings of the 8th Symposium on Dynamic Languages, DLS 2012, pp. 1–16. Association for Computing Machinery, New York (2012). https://doi.org/10.1145/2384577.2384579

38. Purdom, P.: A sentence generator for testing parsers. BIT Numer. Math. **12**(3), 366–375 (1972)

39. Rouş, G., Şerbănută, T.F.: An overview of the K semantic framework. J. Log. Algebraic Program. **79**(6), 397–434 (2010). https://doi.org/10.1016/j.jlap.2010.03.012. Membrane computing and programming

40. The Coq development team: The Coq proof assistant reference manual. LogiCal Project (2004). Version 8.0

41. Wood, G.: Ethereum: a secure decentralised generalised transation ledger (version 2021-04-21). Technical report (2021)

42. Yang, Z., Lei, H.: Lolisa: formal syntax and semantics for a subset of the Solidity programming language in mathematical tool Coq. Math. Probl. Eng. **2020**, 6191537 (2020)

Author Index